material aspects

Exchanging Human Bodily Material:
Rethinking Bodies and Markets

Klaus Hoeyer

Exchanging Human Bodily Material: Rethinking Bodies and Markets

Springer

Klaus Hoeyer
Department of Public Health
University of Copenhagen
Copenhagen, Denmark

ISBN 978-94-007-5263-4   ISBN 978-94-007-5264-1 (eBook)
DOI 10.1007/978-94-007-5264-1
Springer Dordrecht Heidelberg New York London

Library of Congress Control Number: 2012948635

Printed on acid-free paper

Springer is part of Springer Science+Business Media (www.springer.com)

Preface

This book grows out of a longstanding interest in how one might explore "markets for human body parts" in ways that destabilize and explore, rather than presuppose, the meaning of the words "market" and "human body." Bits and pieces taken from patients and other persons are incessantly circulated in a boosting health industry, and they are integral to the advancement of exciting new technologies. These bits and pieces form the basis for a significant quantity of the economy in most high-income countries. Nevertheless, body parts are generally regarded as somehow beyond trade. How might we understand what takes place as people contribute to the development of a highly profitable and booming sector by exchanging something that is not supposed to be owned and used for profit?

Scholars deal with this problem in various ways, but they tend to share certain assumptions regarding what makes a "market" and what constitutes a "body." They rarely explain what makes an entity into a human body part, an object into a commodity, or an exchange system into a market. These categories are presumed, rather than explored. Even if the categories are to some extent shared by scholars participating in the debate, the moral evaluation of the exchanges they describe differs sharply. Either they are morally deplored and the agency involved described as devoid of morality or markets are seen as (morally) superior mechanisms for distribution of scarce resources and heavily advocated to solve (morally) important problems in organ, tissue, and cell procurement. People adopting the first position tend to see existing moral norms as undermined through bodily exchanges, while people advocating the second position typically see existing moral norms as hampering rational allocation systems. Implicitly the analytical gain from the former—the market critical studies—becomes reduced to awareness of "market encroachments" on the integrity of "human bodies," while market proponents tend to deliver a call for market forces to solve problems, which remain focused on the amount of tissue available for recipients at the expense of awareness of a wider set of social implications of tissue exchange. Accordingly, policy options become limited to prohibitions on commercial exchanges of human tissue (based on the assumption that body parts are special and should never be sold) or deregulation of prohibitions on financial incentives (based on assumptions about market forces and a view of body parts as tradable things).

I wish to take another route. I see the prevailing understandings of "markets in human body parts" as constitutive for how such exchanges unfold. I see them as enacting exchanges in particular ways. But I do not take them as adequate analytical descriptions of what is going on. Instead I seek to develop an analytical vocabulary that can help us explore, at a very practical level, what is at stake as various forms of material move in and out of bodies.

I claim that technologies using and engaging with "the human body" challenge ideas about body boundaries at both an ontological and an epistemological level. Body boundaries are not given and never have been—they are under constant establishment, biologically as well as culturally. Therefore, it is an empirical question to explore when and how an entity comes to be viewed and experienced as "part of" a "body." Tissue exchanges interact in productive ways with the making of historically and culturally specific understandings of bodies and their relationship to persons. Likewise, markets constitute a historically specific understanding of exchange. Market thinking is tied to the emergence of particular ideas about what can and cannot be exchanged—and how—and these ideas create special tension in relation to objects moving in and out of the bodily space we identify as persons. By engaging a new vocabulary and using it to unravel some of the social implications of bodily exchange, I hope to have written a book about exchanging the body that avoids reifying its object.

For a decade now, I have studied various versions of what are typically called biobanks, that is, structured collections of various types of tissue with adjoining information about the person in whom the material originates. Biobanks are constructed for research purposes as well as for forensic, diagnostic, and therapeutic purposes, the latter especially in transplantation medicine. The construction of the word biobank has always aroused a certain interest in me: the connection between the biological and the monetary realm. This book is my attempt of getting to grips with the dilemmas faced by ethnographers like myself eager to explore the interconnections between human biological material and monetary interests. I hope it will serve as a resource for future ethnographers traversing the same terrain and being drawn to the same constellation.

The bulk of the book was written during a stay at the Institute for Advanced Study (IAS) at the University of Minnesota thanks to generous funding from the Andrew W. Mellon Foundation. The research preceding it was funded by the Swedish Ethics in Healthcare Programme, the Carlsberg Foundation, and the Velux Foundation, and my travel expenses to Minnesota were funded by a grant from Etly and Jørgen Stjerngren's Foundation. I would like to express sincere gratitude for this support. The IAS proved to be the perfect nurturing ground for a book like this thanks to the marvelous composition of excellent scholars and wonderful support staff. The comfort of this academic environment I have only seen matched by my current workplace in the Department of Public Health, University of Copenhagen. Going to work is a true pleasure when it means meeting the inspiration, support, and good humor of such wonderful people.

With two such workplaces, it seems reckless to call special attention to some colleagues at the expense of others. When I nevertheless do so, it is merely to let everybody know that sometimes I doubt whether the ideas in this book (the good ones at least) are mine or happen to be something I have picked up in conversations with Morten Andreasen, Lene Koch, Henriette Langstrup, Stuart McLean, Sebastian Mohr, Maria Olejaz, Inge Kryger Pedersen, Kelly Quinn, Mette Nordahl Svendsen, Signild Vallgårda, Sarah Wadmann, or Ann Waltner. Lene Koch and Karen-Sue Taussig in particular pushed me (in different ways) to write this book and I appreciate their encouragements. I have presented elements of this book in both departments and benefited immensely from the comments given. Susan Craddock, Jennifer Gunn, Christine L. Marran, Karen-Sue Taussig, and Leigh Turner volunteered for a workshop discussing a sample chapter, and they provided just the needed encouragement and advice. Milena Bister, Anne Carter, Penny Edgell, Henriette Langstrup, Emily Martin, Ingrid Metzler, Reecia Orzeck, Juliana Hu Pegues, and Omise'eke Natasha Tinsley will all find that specific suggestions that they have made have had very direct impact on the manuscript. The result might not be exactly what they meant to suggest, but I am still grateful for the fact that they tried to make me understand what I needed to do. Over the years, discussions with Helen Busby, Kathryn Ehrich, Sarah Franklin, Herbert Gottweis, Maja Horst, Lotte Huniche, Julie Kent, Dorthe Brogård Kristensen, Margaret Lock, Niels Lynöe, Lynn Morgan, Sniff Nexø, Carlos Novas, Naomi Pfeffer, Barbara Prainsack, Silke Schicktanz, Tine Tjørnhøj-Thomsen, Richard Tutton, Lars Ursin, and others have changed how I think about several of the topics in this book, and I wish I could have provided sufficient credit to all of them. A central concept in this book, ubject, was introduced to me by Stefan Helmreich and liberally followed up with further suggestions.

Many people have helped me by introducing me to specific cases relevant to my often bizarre interests. I would like to extend particular thanks for such tips to Hans Okkels Birk, Anne Carter, Louise Kragh Engell, Mette Hartlev, Naomi Hawkins, Jane Kaye, Julie Kent, Karen Kinoshita, Ivan Lind, Naomi Pfeffer, and Tobias Rudolph. Carol Bang-Christensen has helped with appropriate translations of difficult passages from Danish and Swedish to English, Sofie Birk was a great student assistant, and Kirsten Jungersen has helped with Latin phrases.

Elements of text have previously appeared in Hoeyer K (2004) The emergence of an entitlement framework for stored tissue—elements and implications of an escalating conflict in Sweden. Sci Stud 17(2):63–82; Hoeyer K (2005) The role of ethics in commercial genetic research: notes on the notion of commodification. Med Anthropol 24(1):45–70; Hoeyer K (2007) Person, patent and property: a critique of the commodification hypothesis. BioSocieties 2(3):327–348; Hoeyer K (2009) Tradable body parts? How bone and recycled prosthetic devices acquire a price without forming a 'market'. BioSocieties 4(2–3):239–256; Hoeyer K (2010) After novelty: the mundane practices of ensuring a safe and stable supply of bone. Sci Cult 19(2):123–150; Hoeyer K, Nexoe S, Hartlev M, Koch L (2009) Embryonic entitlements: stem cell patenting and the co-production of commodities and personhood. Body Soc 15(1):1–24. I thank the respective journals, as well as my coauthors on one of the pieces, Mette Hartlev, Lene Koch, and Sniff Nexø, for permission to use this material again in a different form and context.

The thoughts of this book are, as I have just admitted, to a great extent communal products of discussions with good colleagues and generous interlocutors. But there is another, more personal, dimension to the writing of this particular book and my work with it. Family and friends and Jesper, more than anybody, continue to remind me of the embodiedness of love and care. Had you, Jesper, not gone along to all the strange places ubjects have taken me, I would never have had the courage or inclination to begin exploring what they mean. Thank you.

Contents

Chapter 1
Introduction

On January 28, 2006, *The Washington Post* reported that "hundreds of very live Americans are walking around with pieces of the wrong dead people inside of them." The macabre news came while the media and police unraveled how a tissue recovery company called Biomedical Tissue Services illegally had harvested bone, tendons, and skin from funeral homes and sold the tissue as implant grafts to hospitals and dental clinics. Informed consent sheets were forged, and papers documenting age, cause of death, and disease history were made up to make the grafts appear safe.[1] The most famous body enrolled in this unwarranted recycling of parts was that of Alistair Cooke, a British broadcaster especially known for his Letter from America program running for 58 years on BBC. He died at the age of 95 in 2004 from a lung cancer that had spread to his bones, which were nevertheless harvested by Biomedical Tissue Services. The horror of what eventually turned out to be more than 20,000 recipients having received implants with fake recovery papers—pieces of the wrong dead people—led many Anglophone newspapers to report on the "tissue-processing industry." In one such report, *The New York Times* noted that the "tissue-processing industry, once limited to whole organs, has evolved quickly as techniques have developed to make use of muscle, bone, tendon and skin in therapies and research."[2] In this way the bones of Alistair Cooke directed attention to something new. However, there is something slightly peculiar about the notion of bone and skin transplants as a recent addition to transplantation medicine: bone has been transplanted for more than a hundred years. Why are bone transplants portrayed as new and why does an old and established medical procedure suddenly become ethically controversial? There is also something intriguing about the notion of "pieces of the wrong dead people." What is the relationship between a piece of bone and a person—and what makes some relationships right and others wrong? When is something inside one body to be considered part of somebody else? Furthermore, the very term "tissue-processing industry" has an effect which makes it different from other "industries." What does the industrial and commercial setup around products that come from bodies imply for the framing of this story and its effect on readers? *The New York Times* might assume that the reason they had not heard about bone transplants before is because it is "new," but we clearly need a

K. Hoeyer, *Exchanging Human Bodily Material: Rethinking Bodies and Markets*,
DOI 10.1007/978-94-007-5264-1_1, © Springer Science+Business Media Dordrecht 2013

more accurate understanding of what is at stake. I suggest that the newness relates to a current reconfiguration of relations between three interrelated domains: the "body," the "person," and the "market." This book is about analyzing this reconfiguration—conceptually, historically, and ethnographically—as a process of change emerging through exchange.

My interest in this reconfiguration of interrelated domains stems from attempts of understanding the stakes in a particular debate which is often bluntly framed as being about "*markets in human body parts.*"[3] It is a debate in which scholars and policymakers tend to position themselves as either for or against a so-called market model for exchange of human biological material. It is also a debate which tends to generate a sense of moral concern: a concern that can also be discerned in the description of the tissue-processing industry allegedly "revealed" by *The New York Times*. The sense of concern that informs the debate is, I suggest, part of shaping the reconfigurations I describe. With this book I aim to facilitate a critical analytical engagement with the well-rehearsed positions in the debate about "markets in human body parts" by way of focusing on their performative effects for the phenomena they claim to describe. After explaining this ambition in more detail, this chapter will outline the common positions in the debate before embarking on a deconstruction of the involved ideas, first about "markets" and what they do (Chap. 2) and then about bodies and what they "are" (Chap. 3). This conceptual work lays the ground for a new empirically grounded understanding of the moral work involved in exchanges of bodily material (Chap. 4), as well as for a more general argument about the ways in which market-like exchange forms reconfigure moralize and humanize bodily material in modern, secular, capitalist societies (Chap. 5), before the argument is tied together and used to point out how a new analytical approach can also facilitate new answers to some of the common queries and controversial questions addressed in the debate about "markets in human body parts" (Conclusion). I elaborate this structure toward the end of this chapter.

The media feature a constant undercurrent of stories relating to how bodies—whole or in part—are shipped around and used for various purposes, stories about what it implies for donors and recipients, and stories about the role of money in facilitating these events. Following the economic crisis beginning with the financial collapse in 2008, gamete bureaus reported how a much greater number of donors suddenly wanted to offer their sperm and eggs.[4] Newspapers continuously write about organ shortage and how poverty forces the world's most destitute people to "sell" their organs.[5] Within the transplant community, which seems to be permanently prone on scarcity, the use of financial incentives is repeatedly up for debate.[6] Furthermore, embryonic stem cells, DNA samples, and other types of human biological material are viewed as essential to what institutions like the Organisation for Economic Co-Operation and Development (OECD) call the "bioeconomy."[7] And at the mundane level of everyday clinical practice, thousands of wards every-day ship around "pieces of people" while figuring out how to organize the monetary aspects of those shipments. Institutions responsible for tissue procurement for research, diagnostic routines, or transplants are expected to develop policies and have to deal with tricky questions relating to the rights of donors, collaborators, and other stakeholders. When trying to figure out who is entitled to what, they face amazingly complex legal issues—issues that nevertheless tend to be presented as very plain: are bodies "beyond trade"? Do body parts constitute property? Who owns them? In the same way, scholars trying to address the immensely complex issues involved often become engaged in debates that take the simplifying shape of "for or against" a market approach to body parts. I will suggest that this very framing is causing more problems than it solves.

Many of the media stories and the organizational quandaries touch upon something which is intuitively important. In their influential book *Tissue Economies*, sociologist Catherine Waldby and literary scholar Robert Mitchell suggest that tissue exchange indicates a special form of social contract: ties, duties, and moral positions can be expressed through literal exchange of tissue.[8] Relations thereby acquire a material form. Other scholars have also shown how tissue tends to be matter that matters: it embodies multiple layers of meaning and its exchange potentially interacts with transformations in the self, issues of identity, and understandings of what bodies are and what can be done to them.[9] Literary scholar Susan Squier says about the constant flow of stories about biological material, which is living what she calls *liminal lives*: "Over our morning coffee we can discover how the foundational categories of human life have become subject to sweeping renegotiation under the impact of contemporary biomedicine and biotechnology."[10] The question is, however, which foundational categories are at stake in the current surge of "tissue economies"? Is all tissue automatically matter that matters? If tissue is important because it is part of a body, what then is a body? Categories like tissue and economy as well as bodies and markets are taken for granted in most of the work analyzing what happens when the domain associated with the body becomes enrolled in various forms of exchange. This is ironic because while the meaning of a word like tissue is certainly in flux, it is implicitly reified in the literature that should describe the changes.

To summarize the ways in which monetary, biological, and scientific values merge in current tissue exchange, Waldby has suggested talking about biovalue as a "surplus value of vitality and instrumental knowledge."[11] There is a long tradition in anthropology for working with broad concepts of value and for emphasizing the way value conveys "ethical, economic, aesthetic, logical, linguistic, and political dimensions of human life."[12] Because tissue is associated with persons and their moral status in ambiguous ways and because tissue is often exchanged within complex medical knowledge systems, we need to think about value in a way that combines moral and epistemological dimensions with the notion of monetary worth, but we must do so reflectively and without presuming too much about what money does and what bodies are. Valuation involves more than negotiation of price.

Though the work on tissue economies is foundational for the terrain that this book covers, I also depart with it in several respects. First, I have problems with the term "tissue" because it seems implicitly to delineate what we compare based on ontological assumptions about bodies as composed of *biological* material of a particular type, a person's tissue. Secondly, I think there is a reason not to depend too much on the economist's conception of value and the economic discipline's basic assumptions about supply/demand mechanisms. Waldby and colleagues have begun to unfold the specificities of what is at stake when we exchange something ambiguously related to human beings, and we need to take this ambition further. To do this, this book argues, we need to move beyond the vocabulary of tissue/body and market/economy. Since the abolishment of slavery, Western legal systems have sought to uphold a legal binary between persons and commodities. The binary is sometimes challenged, for example, in court cases revolving around the products of new medical technologies such as cell lines, embryos, or gametes. In fact, as a consequence of a growing use of human biological material in the biotech industry, the legal system repeatedly has to deal with something in between persons and commodities, though poorly equipped to do so.[13] Similar ambiguities are at play when courts are engaged in, for example, determining who may be fertilized with deep-frozen semen kept in storage from dead men; who may have children using embryos stored prior to a divorce; who can inherit the ashes of a dead parent; who may control the uses of a removed tumor; who may conduct research on stored blood samples; and whether removal of organs, tissue samples, fingers and ears, and even metal remains from hospitals, morgues, and crematoria constitutes theft or should be punished according to other legal doctrines.

In short, when something ambiguously related to the bodies of persons becomes exchanged, it has consequences for a wide range of values and concerns, and to explore them we need a fresh analytical approach positioning itself differently in relation to the "foundational categories" at stake. We need to move beyond "body part," "market," and "property rights" to study what these categories *do* to the exchanges. In the following I unfold the vocabulary I suggest, its wider theoretical framework, and the nature of my analytical ambition, before I outline the positions in this intriguing debate about "markets in human body parts."

Making the Case for a New Vocabulary: Exchange, Ubject, and Entitlement

First, the term "market" is deeply problematic for several reasons. When used in conjunction with "body parts," it instigates strong moral reactions which tend to seriously impair empirical analysis. Also, it is associated with theories that are not necessarily helpful for understanding how material originating in bodies moves through so-called supply/demand mechanisms, and to avoid this package of ideas, I prefer focusing on the actual movement of this material, that is, its *exchange*. By focusing on systems of exchange in my analytical vocabulary, I can keep it as an open question when exchanges are seen as "market-like" by particular actors, and I can explore what such perceptions do to the exchange rather than settling it simply by way of my wording. Chapter 2 will unfold the reasoning behind this argument in detail.

Secondly, is anything passing through the space that can be said to be phenomenologically experienced as body also a body part? For how long does it remain a body part after having left that space? Who is it part of after a transplant? And how do you distinguish between the object in question and the subject making the claims? The words we use to pose these questions intervene in our handling of the involved ambiguities. If you write "object" rather than "subject," you have already applied a conceptualization which implicitly facilitates a property discourse, but this very conceptualization is at the heart of what many actors contest.[14] To avoid simply having the categories doing the thinking for us, an alternative vocabulary is therefore necessary. Material, which is ambiguously related to bodies and to persons, is neither fully subject nor fully object, and to capture this basic ambiguity, I suggest using the word *ubject*. Over the years I have considered many other terms,[15] and of course I have also tried to avoid the use of neologisms altogether. But I have come to realize that a new term is needed for this particular argument, simply because the debate that I analyze has come to an impasse as a result of far too many implicit assumptions attached to the existing vocabulary of body parts. "Ubject" is a strange word which will appear odd (or even annoying) to most readers, but it facilitates a position from which the productivity of the ambiguities so central to this field can be explored.[16] I do not expect or encourage others to use this word in their own writing. On the contrary, it is intended to provoke readers by way of being alien. I wish to provoke readers to rethink their own practices of categorization when subsequently using these terms, not to adopt a new prefixed category. By way of being alien, strange, and even annoying, the word ubject draws attention to the basic ambiguities surrounding the materials floating through bodies. Had I chosen to write body part, tissue, or commodity, the "real" nature of the exchanged entity would have been settled through my naming and framing.

Ubject, for me, designates that which is seen by (at least) some as having been part of a body and therefore related to a subject. The subject-object divide is closely associated with other distinctions—persons versus things, owners versus commodities—and these associated distinctions strongly shape how we construe agency, rights, and entitlements. By means of an analytical vocabulary that does not employ the distinction, it is easier to study when and how the distinction between object and subject becomes active, what it does, and what reactions potential transgressions might generate. Furthermore, if it is not clear what is "part of a body" (and whose body), we might learn something important by posing this as a question rather than using a category like tissue which implicitly sets up boundaries for "proper body parts." With ubjects we can explore a wide range of flows through the space typically experienced as body (instead of limiting ourselves to "tissue") and see when and how an ubject acquires the status of a body part and what this status does to its mode of exchange. We need not presuppose that people treat ubjects in special ways thanks to a particular biological heritage. Implants and devices, for example, form part of bodies at various points in time, and, in fact, this book will show how exchange of metal parts that have passed through "bodies" in surprising ways can resemble exchange of certain tissue types. To explore the basic ambiguities surrounding the category of body, we therefore need an open category. Chapter 3 unfolds the reasoning behind this argument in greater detail.

Thirdly, it is necessary to think about the words we use to discuss the relationship between legal persons and ubjects. The word property is used to refer both to a feature of a thing and a relationship to a thing. While objects can potentially be owned, subjects are not supposed to be possessed, and an idiom of ownership implicitly designates ubjects as passive, something potentially possessed but not possessing. Furthermore, ownership is for many people closely affiliated with a potential for sale, commodity status, and characteristics associated with the pervasive institution referred to as "the market." It is not at all clear, however, what type of rights and duties ubjects engage, and often great care is expressed to make sure it does not appear like "markets" when ubjects are exchanged. Hence, ownership and property are part of the empirical struggle to settle what ubjects are and how they may be treated, and are therefore unfit as analytical concepts. The term "ownership" is extremely vague. Legal textbooks, for example, give no clear definitions, but it remains modeled on a distinction between persons and things. Ownership and property rights are often used interchangeably, meaning a "bundle of rights allowing one to use, manage and enjoy property, including the right to convey it to others."[17] The conceptual vagueness is replicated in a lack of legal clarity about the rights persons can acquire toward "body parts."[18] Some legal scholars find the vagueness of the current legal situation unsatisfactory and suggest employing new categories, for example, cyborg, on which to build clear rights to ambiguous entities.[19] Irrespective of your legal ambitions, however, the point is that to understand this type of negotiation, you need a vocabulary that avoids reifying the rights people aspire for in ways that settle the basic ambiguity about the status in between person and thing. I think *entitlement* is a good term. While there are various legal theories concerning how property regimes emerge and the purposes they serve, they all associate ownership with power and legitimacy.[20] In other words, we can view ownership as a particular form of entitlement backed by a specific historical development and a legal system employing particular distinctions.[21]

Relational Concepts in a Moral Landscape

Entitlements operate in a moral landscape. They must possess a certain degree of moral legitimacy to become established as de facto rights and not just potential legal rights.[22] It is not uncommon for social scientists to privilege power in their analysis of ubject exchange, but it is absolutely essential not to analytically subordinate morality to power if we wish to pay tribute to the subtleties of human agency. Notions of what one *ought* to do—even when it is not in one's immediate personal interests—are important for most human beings. I insist on using "morality" (noun) and "moral" (adjective) as analytical concepts when something is at stake for people, and they are willing to do something for it, though they do not see it as in line with their personal, more selfish, interests. Moral notions are not external to the games of power and truth and cannot be reduced to an effect of them either.[23] An analysis that uses concepts like property and markets as factual descriptions is often blind to the moral agency going into making the exchange of ubjects different from exchange of, for example, light bulbs or vacuum cleaners. And it implicitly overemphasizes the monetary aspect of the exchange at the expense of the other types of values and concerns associated with ubject exchange.

Therefore, I suggest studying *entitlements* and *exchange systems* rather than property rights and markets and to say *ubject* where others might have said body part or prosthetic device. Basically, the point is that if we wish to understand what is at stake in the controversies revolving around "markets in human body parts," we first need to be able to ask when and how something comes to be seen as a body part, when and how entitlements become construed as property rights or ownership, and when and how exchange systems are enacted in the image of the free market. And we need to remain open to ubject exchange having implications other than monetary value generation—or what is simply called profit.

I think of the three central analytical terms as referring to relations rather than entities. "Ubject" is a *temporal relation* (a step on the way from having been part of a body to not being so anymore) and not an entity (though of course I cannot change the fact that when we talk about materials flowing through bodies, we tend to talk about something already conceptualized as entities, as it is also the case with alternative words such as "body part," "tissue," and "commodity"). The strangeness of the word ubject is designed specifically to remind readers that ubjects change status over time as it moves between subjecthood and objecthood. Entitlement is a *relationship of power* and not a name for a defined set of rights. Exchange is a *social relation* enacted in concrete practices and not a delineated realm of the world as some see markets. With these words I can better explore the basic questions we need in order to explore "markets in human body parts," namely, "What is a body?" and "What is a market?"

Asking such questions might appear strangely ontological. Do we need to know what bodies or markets are? No. And I do not claim to answer them. I pose these questions to illustrate that no easy answers apply. Unless directly addressed implicit assumptions about bodies and markets remain tacit and the work they do on our thinking remains unacknowledged. Ontological presumptions always lurk around somewhere in the background[24]—Why not make them analytically active? Markets and bodies are not given entities waiting to be discovered; they acquire their meaning through specific forms of interaction.

The Wider Analytical Framework

I approach the influence of market thinking on ubject exchange as an element of a particular biopolitical configuration shaping available spaces for action.[25] In his late work on biopolitics, Michel Foucault construed a particularly apt theoretical point of departure for studying the organization of ubject exchange.[26] Focusing on the triangulation of truth, power, and ethics, he pointed to power as the name of a complex strategic situation in which people come to view themselves and their

'tissue' meaningless –
Continual reconfigurations

options in particular ways. This triangulation fits well with an interest in biovalue—the interrelated moral, monetary, and truth value—that Waldby has identified in many ubjects.[27] With his biopolitical framework, Foucault traced the emergence of phenomena that have otherwise come to be taken for granted. Instead of presupposing particular universal entities and projecting them back through history, he focused on continual reconfigurations of phenomena over time. It is an analytical predisposition which has inspired much of my work, and it runs as an undercurrent throughout this book.[28] Foucault describes power as relational, positive, and decentered. Nobody holds it but still it shapes, produces, and frames the problems that become most compelling and the solutions that appear most appealing and legitimate at a given moment. Ownership constitutes a form of power in the Foucauldian sense of the word as a relational, decentred, and productive force in the world: a structuring principle respected and upheld by both dominating and domineered, by the haves and the have-nots. I do not think of power in an agential manner as if it did or wanted something[29] and refer to biopolitics as much as a product of agency as a precondition for it. Nobody and nothing can act out of context, and context involves an intricate and indeterminate interplay of truth, morality, and relative strength. Biopolitics denotes a given strategic situation in which this intricate interplay performs and shapes the problems that a given society addresses.[30]

I pay relatively little attention to concrete policies[31] and more to logics and mechanisms involved in the biopolitical situation through which the debate manifests itself. Some see biopolitics as too governmental in the connotations it produces. The strategic situation in which current problematizations unfold is, they argue, more fittingly described as a bioeconomy.[32] I am intrigued by these analytical developments and not least by the ways in which they have reintroduced a Marxian perspective on exchange. For the reasons described above, however, and to avoid sounding like an OECD policy paper, I wish to keep away from "economy" as a primary analytical concept. What the bioeconomy literature has delivered, and that we need to appreciate, is a critique of a particular reception of the Foucauldian approach which has foregrounded discourse and meaning at the expense of understanding the material aspects of exchange. Analysis of the mode of production and distribution is a necessary step toward understanding the strategic situation I describe as biopolitical, but one should avoid choosing between "basis" and "superstructure" as the preeminent force from which everything emanates. Societal discourses and modes of production are coproduced as argued in more depth in Chap. 2.

Bodies and Persons

The latter point goes also for ideas about bodies and ubjects: their material and semantic lives cannot be separated and understood apart. During the past few decades, many interesting things have been written about bodies and "body parts." The so-called new technologies of the body are often charged with causing a mess that society must try to clear up, but this just might be accusing new technologies for more than they should bear. Paul Rabinow has suggested that it is not so much new technologies as it is very old ideas about the relationship between bodies and persons that are causing a mess. In particular, a notion of *material continuity* between body and person stirs the waters in which ubjects move. Rabinow even suggests that "The intimate linkage between the two key symbolic arenas, 'the body' and 'the person', would have to figure prominently on any list of distinctively Western traits."[33] Rabinow builds his point on the seminal work by Caroline Walker Bynum on the permanence of ideas about material continuity from medieval theology to current debates about organ transplants. Bynum makes the bold assertion that

> …it is certainly today true that considerations of self and survival take the body with impassioned seriousness. We face utterly different problems from the schoolmen and artists of the Middle Ages. Yet the deep anxiety we feel about artificial intelligence and organ transplants, about the proper care of cadavers, about the definition of death—an anxiety revealed in the images of bodily partition and reassemblage that proliferate in our movies and pulp fiction—connects us more closely than most of us are aware to a long Western tradition of abstruse discussion of bodily resurrection.[34]

While Bynum is no doubt right that we are more closely connected to medieval thought on this topic than usually believed, it is less certain that it is the *same* concerns disturbing current ubject use in medical technologies. In fact, I wish to argue that the current obsession with body boundaries cannot be attributed solely to resilient ideas about material continuity between bodies and persons, just as it is too narrow to attribute the current obsession with the body to new technologies producing hybrids never seen before as other scholars have argued. Current semantic battles indicate a changed configuration of bodies and persons in relation to the institution known as the market and the material forms of production and exchange it is usually said to denote. To understand what is at stake, we must study the actual flows of ubjects and the concerns they cause. This involves leaving the purely semantic level as well as the "economic" logic to see what ubjects *do* rather than what they *mean* as they circulate, accumulate, and/or dissolve. I wish to suggest that it even involves taking a phenomenological perspective into account to get a grasp of how bodies feel and what makes some ubjects more likely to be experienced as part of a body than others. I therefore move between worldviews (a semantic cognitive understanding) communicated through shared categories and the lifeworld (the phenomenological experience of being-in-the-world) to learn from the way in which being a body often effaces ready-made explanations of what such a body is and what is part of it.[35] The point is not to determine what bodies *are* but to understand better—to borrow an expression from Annemarie Mol—how they are *done*.[36]

In this *doing*, discourse and materiality are coproduced. In *We Have Never Been Modern*, Bruno Latour describes a particular modern way of thinking which emphasizes separation of the realm of humans (discourse) from that of things (materiality). This thinking involves two types of work, that of purification (making dichotomies) and that of translation (transgressing them with hybrids). Hybrids are unruly semantico-material transgressions of categories. Latour's argument is that hybrids *precondition* the work of purification. This is instructive for an understanding of ubjects. Ubjects in many ways represent unruly semantico-material category transgressions.

If ubjects are hybrids in Latour's terms, they are parasites in Michel Serres'. For Serres the parasite (the French term also confers noise) is that which evades units. The parasite might be subversive, but it is not destructive. The parasite produces the images of that which is clean, much like Latour's hybrids produce purification.[37] Following this line of thought, we can think of ubjects as hybrids or parasites instigating new semantic battles through which conceptions of persons and bodies as well as, gifts and commodities are purified. They instigate what Herbert Gottweis and colleagues call ontopolitics understood as "political contestations in which questions about 'what to do with an entity' are related to or translated into struggles and debates on their categorization."[38] Ambiguous and unruly, ubjects often awaken uncanny feelings. In a famous essay, Freud suggested that the uncanny is related to the revelation of something which ought to have remained hidden,[39] something that makes a bad fit for how we prefer seeing the world. Not unlike Latour, Freud suggests that uncanny transgressions are as common in a modern society employing ideals of clear and transparent categories as they are in so-called primitive societies (irrespective of the prevalence of such ideals). Freud, however, links the uncanny to repression and a particular set of ideas about a universal psychology to which I do not subscribe. I prefer the emphasis in Latour, Serres, and also Foucault on what the uncanny does, that is, its *productivity*, rather than what it is supposed to repress.

The analytical perspective that I have chosen provides opportunities and obligations. How do I choose ubjects that do not privilege particular notions of body? Or types of exchange that are already market-like or unlike? While drawing on many different examples from the literature, I will use Chap. 4 to focus on three types of ubjects characterized by being circulated in forms of exchange rarely discussed in relation to each other: blood samples used for high-tech commercial genetic biobank research, bone used for low-tech therapeutic transplants, and crematoria recycling. In other chapters I compare these ubjects with well-known examples of other ubjects. My point is of course that various ubjects have something in common and that it makes sense to construe a comparison of their forms of exchange, but also that we should take care not to extrapolate too much from examples picked using implicit categorical choices such as "tissue," "organs," "gametes," or simply "body parts."

Wetlands: Clarifying the Analytical Ambition

I need to do a little more to clarify what I want this book to do and who it is written for. I take point of departure in an empirical problem, ubject exchange, and a cross-disciplinary scholarly debate, "markets in human body parts," but this book has as its primary audience the community of anthropological, sociological, and STS scholars who like myself explore ubject exchange ethnographically and yet engage in debates taken by philosophical ethicists, economists, and legal scholars. I hope that scholars from these other disciplines also will find it useful in helping them to approach debates about ubject exchange from new angles, but I also know that for some it will demand a little more work than for those traversed in the literature I primarily employ. The book is motivated by four interrelated analytical ambitions.

They simultaneously represent my contribution to the biopolitical literature that I am inspired by and aim to develop further.

Firstly, I have already indicated how I think we need to pay more attention to moral agency. I find it important to remember that acts stimulated by nonselfish concerns can have cruel effects. Moral agency is not about doing good; it is about *doing* something because it seems important even when the agent does not stand to win or lose anything from it as a personal stakeholder. Moral agency is seen by many market proponents as related to "superstition" or "cultural barriers to a rational approach," while many market opponents see it as representing "sane resistance" to "cynical markets." I find both approaches inadequate and the book is composed to show why. utilitarianism

Secondly, I wish to integrate the work on high-tech, frontier technologies with the mundane everyday handling of body parts. I think much of the literature places far too much emphasis on newness and draws too many conclusions concerning what is just about to change or what is a consequence of the absolute latest scientific achievement without considering well-established, routinized, and relatively uncontroversial ubject flows. This book aims to place more emphasis on the mundane, pervasive, and taken-for-granted technologies of everyday life in the field of ubject exchange.

Thirdly, I argue the relevance of synthesis and comparison to a field in which most scholars emphasize localness and ethnographic grounding and context and squirm back from anything which could read as a sweeping ontological claim about the world in general. Universal categories stand as a capital sin in the tradition that I see myself as drawing on and tapping into, and some of the old comparative work did indeed aim at such universal categories. Nevertheless, I think there is a case to be made for returning to elements of this comparative effort; only we may compare without pretending to compare universals.[40] Marilyn Strathern points out how we make partial connections between unrelated phenomena in the course of our study: it is a way of knowing.[41] I use ubjects to make such connections—I use it to facilitate uncommon comparisons—but my point is not to make a grand theory about ubjects as a universal phenomenon. I think of theory building as case specific. I have constructed a particular vocabulary to create a new angle with which to address a particular debate in a deadlock. I do not see it as a step toward a theory about the world in general. With small adjustments, the notion of ubject could come to mean other things and address other problems, but this is not my ambition. Others would want to explore what is ambiguously related to persons on other fuzzy boundaries – human/animal, person/representation, worldly/spiritual, or normal/monstrous – or they might want to explore cases where the subjectivity of living persons is questioned in one way or another (e.g., certain cases of research subjects, trafficked women, adoptees). Of course, people may use the ideas presented here as they see fit to address such topics, but this book is not about all such potential contestations of subject/object status. It is about understanding reconfigurations of three interrelated domains: the "body," the "person," and the "market" enacted through debates tagged as revolving around "markets in human body parts."

To be able to capture the basic ambiguity surrounding something which both is and is not part of a body is the fourth analytical ambition, and it demands a little further explanation. The central analytical figure guiding this book is the productivity of the undefined. I think of the undefined as a *wetland* with no stable grounding.[42] This interest in what ambiguity does explains my choice of vocabulary and most of the examples I employ. I accept the attraction of dry spots of clarity, understood as clearly defined positions or familiar concepts that we easily recognize such as a pro- or antimarket argument or the notion of a bounded body with well-defined parts. Personally, however, I prefer to move in the wetland of in-betweenness. Many ethnographers probably view themselves in this way, and some also think that they pay tribute to ambiguity when they write about "body parts treated as commodities." I disagree. I believe they are lured by their own vocabulary to construe a sense of moral concern stimulated by a sense of illegitimate transgression. Their texts produce a sense of longing for dry spots (a longing for purification work) so that bodies and the world of commodities can be safely divided. My dedication to wetlands goes beyond the particular debate. I open each of the following four chapters with identification of two dry spots in existing literatures to clarify what type of wetland, in between, I set out to traverse. I also move between genres and argumentative styles used in legal studies, ethics, science and technology studies, and anthropology and thereby insist on taking residence in a disciplinary wetland where, I guess, nobody really feels at home. I like this type of discomfort.

Ambiguity is not only an analytical ambition relating to concepts and theoretical or disciplinary positioning; it is also a moral and political engagement. Many scholars make laudable attempts of *solving* the issues at stake in ubject exchange. I appreciate and acknowledge their dedication, but it is not my ambition, even though in the conclusion I do suggest some alternative ways of framing problems at stake. Every proposed solution involves a set of blind angles, and following Foucault I see my own primary role in being on the move to expose the blind angles, of existing solutions. It might sound tiresome, but at least you are never out of work: "If everything is dangerous, then we have always something to do,"[43] as Foucault puts it. The people taking strong positions in the debate about "markets in human body parts" always seem to know what the main dangers are and how to avoid them, but anthropologist Michael Jackson rightly points out that passing a moral judgment is usually part of closing a case.[44] I prefer roaming around in the morally ambiguous wetlands a little longer than what is usually considered good taste. Friedrich Nietzsche claimed that "a philosopher has nothing less than a right to 'bad character.'"[45] I am not a philosopher, but I too occasionally depart from the safe company of the righteous. I think we can all learn more about the current dangers if we are not too predisposed to see particular dangers already defined through the categories and discursive formations available to us. Hence, my aim is not to "correct moral flaws in current thinking" but to expose what the current thinking does, so that both market proponents and opponents might reassess the issues at stake in ubject exchange in light of their own hopes, aspirations, fears, and concerns. More specifically, I hope this reflective work will help in producing more nuanced ethnographies.

Before I go deeper into the wetland, I will outline the dominant dry spots in the debate about "markets in human body parts." I take a deliberately arrogant position toward the otherwise thoughtful and interesting contributions to the debate. Rather than emphasizing the usefulness of the insights reached by all these contemplative scholars, I am driven by a curiosity of a different kind. I am interested in the blind angles of existing debates, in the logics that make particular arguments and concerns prevail at the expense of others, in the ethos and morality implicitly expressed *through* debates rather than the positions as such taken *in* the debate.

Market or Gift Economy? Outlining the Debate About "Markets in Human Body Parts"

As scholars make themselves into proponents of particular ways of organizing the exchange, positions tend to fall somewhere in the continuum between two models: at one end of the spectrum, the ubject is viewed as (ideally) a commodity, entitlements as property relations, and exchange systems as (unfortunately imperfect) markets, and at the other end, the ubject is viewed as part of a subject which should become detached and exchanged only as a gift, where entitlements relate to information, dignity, and medical needs, but not money, and the exchange systems constitute a gift economy motivated by altruism rather than self-interest. A market model provides donors (or their relatives) with full-blown property rights and a right to profit; a gift model limits the entitlements of donors to informed consent and dignitary interests and sees itself as conserving the "true value" of the ubject.

When people try to persuade their readers that a model is particularly expedient, appropriate, or suitable, it is not always clear for *whom* or according to which *criteria* this assessment of efficiency, appropriateness, or suitability is made. The first step of analysis is therefore to look into this issue. This is how we might begin to understand the implicit ethos and morality expressed through the debate and the expectations scholars hold to exchanges framed as markets.

Pro-market Arguments

Pro-market arguments mostly focus on ensuring a stable supply and an efficient system of allocation and sometimes on ensuring donors a share of potential monetary benefits. The typical pro-market arguments will claim that a market already exists: "The fact is that there *is* a market, and to say that there is not is to perpetuate a fiction"[46] This "market," however, is seen as hampered by obstacles—it is imperfect and without proper supply/demand mechanisms. If these obstacles were removed, more people would be willing to sell organs, tissue, or cells which would result in an optimization of health. Whose health? Well, the persons who need the ubjects— and who are in a position to pay.[47]

Some market proponents emphasize the philosophical and legal sides of the issues and state that to be owned, the ubject must be a thing because property "expresses the rights of persons in and over things."[48] This need not be a problem, Rohan Hardcastle argues, because simple detachment from a body makes any part into a thing.[49] Hardcastle does not inform his readers about the potential limits as to how much that can be detached without the part becoming a whole. Other legal scholars are more aware of this type of problem of defining what a body is.[50] Mason and Laurie, for example, propose a property model but acknowledge that it cannot be absolutely clear what forms part of the possessed body and how much can be removed and still be considered only things. During an autopsy, blood and tissue bits will be spilled, for example, but they suggest that if one employs a property approach, "each possessor is entitled to find the body in the state he or she antici-pated," and clearly, they find, it would be unreasonable to expect the return of a body including *all* blood and tissue pieces.[51] They contend that a legal assessment of "rea-sonable expectations" will thereby settle the ambiguity. The prime reason for them to propose property rights to donors is to provide them with more leverage in relation to researchers and industry. Mason and Laurie claim that informed consent only gives control until the point of donation, while a property model could extend donor control beyond that point. And it would be more just. Like several other authors, they are concerned about a state of affairs where everybody but the donor seems to make money on the exchanged ubjects. They see the provision of property rights to donors as a way of providing them with tools of power.[52] A strong interest in justice and ensuring the rights of donors also underlies other pro-market arguments. Michele Goodwin wishes to show solidarity with the black Americans who typically stay the longest on waiting lists and who have been abused in so-called altruistic procurement systems. She believes that a right to trade is the proper way to reach that aim.[53] With a similar view on justice, Cécile Fabre argues that bodily goods are akin to other goods and like other property items body parts should be subject to taxation and obligations of sharing.[54] Proper control with markets—including markets for body parts—is better and fairer than to claim there is no (monetary) value of the desired good.[55] Mark Cherry goes further and proposes that only by installing a right to sell, society would be showing respect for the autonomy of kidney vendors:

> While prohibition is robustly paternalistic, demeaning the poor and sick by considering them unable to make moral decisions about their own fates, the market respects the dignity of vendors and patients as persons and rational moral agents able to make choices about their own lives.[56]

Most pro-market arguments focus on the interests of the *recipients* rather than the donors, however, and rhetorically ask why people must die on waiting lists just to uphold strange ideas about the ethical status of body parts.[57] Often, the register used to pose this type of question is highly indignant as when Sally Satel decries the "hundreds of thousands of needless deaths"[58] easily avoided with a market model or when James Stacey Taylor concludes "there is a simple solution" to all the suffering, namely, "legal trade in human organs … [the] implementation [of which] is morally imperative."[59] Arguments against a market position are dismissed as "virtually with-out exception, illogical, unfounded, or highly speculative" or "at best, suspect, at worst, nonsense."[60] Usually the current situation is seen as resulting from irrational behavior among politicians,[61] and some specifically blame the culture of the "organ transplant establishment."[62] A few economists, however, refrain from ascribing such power to culture and suggest that the transplant community must stand to gain financially from the current predicament since only (monetary) self-interest can explain people's attitudes.[63] The rational alternative would, according to these scholars, be to rely on "market forces" and the "invisible hand of the market"— using Adam Smith's classical metaphor.[64] The invisible hand would eradicate the shortage of organs, and thus the "needless deaths," by way of offering a price that would make either living donors or the relatives of the deceased provide the desired ubjects. A shortage is, in this line of thinking, just a matter of price: "A shortage is, by definition, an excess of quantity demanded over quantity supplied at the prevailing price. As such, a shortage is a distinctly economic phenomenon."[65]

Some pro-market scholars believe that opponents are mistaken when fearing that prices will be very low (and thereby only add misery to the fate of the poorest in society): they believe prices will rise and provide sellers with enhanced autonomy because they now possess cash with which to fulfill their desires.[66] Others think that if "living kidney become more readily available, this price should eventually decrease, leaving buyers with a surplus."[67] Despite disagreeing on the fluctuation in price (depending on whose interests they foreground), they seem to agree that if only prohibitions on trade are removed, some type of "market force" will take care of whatever they perceive as the main problem. Some accept that various forms of emotional attachment might continue to be an impediment for some potential vendors, and yet this attachment is mostly construed as a manageable "opportunity cost." As two economists pose it:

> We expect that the opportunity cost of supplying the organs is rather low. Generally organ removal causes no visible disconfiguration to the body (indeed open-casket funerals can still be held), and the only apparent alternative use is to bury the cadaver with the organs in place. Or, as some in the transplant community have graphically stated it, to "feed them to the worms".[68]

In response to market opponents who suggest that body parts are essentially different from commodity objects, it is said that such arguments "must be attributed to a certain level of naiveté or outright ignorance concerning the normal operation of market forces."[69]

Other pro-market arguments revolve around clarification of the property status after the point of procurement in order to ensure efficient market mechanisms in the types of research that are dependent on access to ubjects. They tend to focus on the work added to the tissue and suggest developments in statutes regulating intellectual property rights (IPR) to straighten out the remaining ambiguities.[70] For these scholars, solidarity with the researchers (and the people able to buy their products) seems to influence their assessment of a suitable solution, unlike the scholars focusing on increased organ supplies, who mostly build on solidarity with recipients in need (and with suitable income levels). Scholars focusing on donor entitlements to profit conversely build their analysis on solidarity with particular types of donors who are in a position to say no to a bad bargain. Each position is associated in this way with particular views of motivation and forms of solidarity, and depending on the position taken, property rights and markets are expected to perform different tasks. In this way, every paper is infused with a set of goals and ideas about "what markets do." More or less the same can be said about the arguments against a market approach.

Against Markets

As we shift from pro-market arguments to arguments against market models, we simultaneously shift from a mainly utilitarian mode of argumentation (focusing on outcome) to arguments more likely to draw on deontological ethical positions. In particular, the German philosopher Immanuel Kant's understanding of human dignity is an important source of inspiration for market opponents, though not all opposition is related to his philosophy. In fact, resistance to market models is also associated with solidarity with the donor and concern for the interests of the recipient—just as we saw it with proponents of market models. Only it is a different type of donor that scholars imagine as point of departure for their arguments, and the interests of recipients are construed differently. All three types of oppositions merge in various mixes in what I will describe below as a commoditization hypothesis.

Whereas most market proponents praise the right to autonomy—including the right to sell one's body parts—the Kantian perception of autonomy is related to a different logic. Kant emphasizes a *duty* to respect human dignity. For him autonomy is not a carte blanche to do as you please, but a source of obligation to restrain your doing. And among the prime obligations is to treat others as well as yourself as an end rather than a means. On the topic of property rights in one's own body, Kant writes:

> Man cannot dispose over himself, because he is not a thing. He is not his own property—that would be a contradiction; for so far as he is a person, he is a subject, who can have ownership of other things ... for it is impossible, of course, to be at once a thing and a person, a proprietor and a property at the same time.[71]

He continues:

> He is not entitled to sell a tooth, or any of his members ... Human beings have no right, therefore, to hand themselves over for profit, as things for another's use in satisfying the sexual impulse; for in that case their humanity is in danger of being used by anyone as a thing, an instrument for the satisfaction of inclination... The moral ground for so holding is that man is not his own property, and cannot do as he pleases with his body; for since the body belongs to the self, it constitutes, in conjunction with that, a person; but now one cannot make one's person a thing, though this is what happens in vaga libido.[72]

From a Kantian perspective, dignity is inherent to a human person and must be defended. Some of his followers even today see the duty to respect this dignitary principle as falling on the "seller" rather than the "buyer,"[73] though one might say that it takes two to demean dignity through trade. By way of a gift model and by limiting entitlements to informed consent rather than profit, donor's dignitary interests are seen as better conserved.[74]

Other market opponents wish to protect the donor, not against himself or herself, but against market exploitation. They are less interested in dignity and more worried about the harm stemming from a world of unequal opportunities where the poor feel forced to give up their organs and body parts to wealthy recipients. In contrast to pro-market version of donor solidarity, trade is here seen as a potential cause of bodily harm rather than an opportunity for profit.[75] This is because the solidarity is invoked with a different type of donor in mind. The pro-market conception of the donor was a person in a position to decline offers of no personal interest. The market opponents, in contrast, imagine donors who are too poor to decline offers of even the worst nature. This donor type is the one we already see enrolled in illegal organ transfers for money in countries like India, Pakistan, and Egypt. From a feminist perspective, Donna Dickenson has argued that women more than men need protection against exploitation because their bodies are more likely to be reduced to passively traded objects.[76]

A quite different argument for a system based on gifting rather than trade takes point of departure in the interests of the recipient. Whereas solidarity with recipients among market supporters focused on the quantity of available ubjects, opponents to market exchange tend to focus on the quality of the available organs. If people donate for the wrong reasons, the supply of organs, tissues, and cells will be characterized by increased risk of disease transmission. The classical argument along these lines was posed by Richard Titmuss in his book *The Gift Relationship* where he compared American (primarily for profit) and British (nonprofit) blood transfusion systems.[77] He found a higher contamination rate in for-profit organizations and explained it with donors being motivated by money rather than care for recipients. Titmuss' arguments have been criticized on numerous occasions, not least following the fatal HIV infection of large quantities of donor blood during the 1980s at a time when gay men were overrepresented among nonprofit organizations' donors.[78] *Healy* Nevertheless, it is the need for reliable donor information that serves as one of the prime arguments for the current global shift in policies toward "gift models" in blood and tissue procurement.[79] The gift model is for these scholars thereby only superficially related to dignitary interests and donor solidarity; it is in fact primarily founded on solidarity with recipients and their needs. These needs are seen as related to the *quality* of the ubject (instead of the *quantity* of ubjects highlighted in the pro-market argument above) and assumptions about procurement models where the monetary incentives so central to the supply-demand mechanisms revered by market proponents are seen as a source of fatal error rather than a wanted instrument to increase the supply.

A very dominant form of criticism of market models among philosophers and social scientists alike can be summarized as the *commoditization hypothesis*. It conjures elements of the above-mentioned sources of critique, in particular the arguments related to dignity and exploitation. Commoditization—or commodification—is a process of disaggregation and marketization of what ought to have remained an inalienable whole.[80] Many scholars are like Donna Dickenson repelled by the way new medical technologies "disaggregate the body, robbing it of its organic unity and encouraging the view of body parts as separate components which do not sum to anything more than their compilation."[81] Because of the prominence of this type of criticism, I will go into a little more depth with what commoditization is seen to imply.

In most cases authors use the concept without specifying its specific theoretical bearing.[82] There is one point, however, about which the reader is never in doubt: it is clearly negatively charged. When Suzanne Holland explains that "commoditization contributes to a diminishing sense of human personhood,"[83] she indicates the nature of the criticism: if an object of human dignity becomes tradable, it is transformed into something of lower moral standing. But how are the dynamics of the process construed? From a philosophical perspective, Wilkinson has suggested that commoditization involves an attitude toward human beings and body parts which (1) denies their subjectivity, (2) prescribes instrumentality, and (3) facilitates exchangeability.[84] Logically speaking, money need not be involved in this definition, as Wilkinson takes care to point out. In the social science literature, however, it seems often to be exactly monetary aspects and market-like exchange forms which give rise to accusations of commoditization.[85] Lesley Sharp is particularly well known for her work on "the increased commodification of the human body and its parts."[86] In one of her books, she takes special care to explicate the concept of commoditization, and I will use this book to elucidate what commoditization is typically seen to involve for social scientists.[87] Sharp claims that commoditization is

> ...a trend that figures in many chapters of human history and cultural settings but has become an especially pervasive—not to mention pernicious—force specifically within the United States in the late twentieth and early twenty-first centuries.

This force takes the form of an *ideology*, which *hides* and *denies* the process of commoditization. It is countered, however, by donors and recipients (i.e., "lay" people), who exhibit "...highly creative ways to confront grief as well as subvert the medical commodification of the human body"[88] through a series of "subversive social responses that ultimately challenge the mystification of commodified organs' origins."[89] Sharp's examples often highlight the monetary dimensions of organ transfers, and she seems mainly concerned about the *moral* dimensions of alienating body parts as infringements of integrity. Another well-known anthropologist, Nancy Scheper-Hughes, emphasizes the *political* dimension of the infringement by pointing to the exploitive character of the alienation of body parts.[90] Other scholars have criticized not only the commoditization of body parts but also the "commodification of bioinformation"[91] also in the case of biobanking and patenting of human genes in general.[92]

The notion of commoditization rests on a number of assumptions about persons and commodities which are in many instances shared with the market proponents. The emphasis placed on the key distinction between persons and commodities is a historically specific product, as I will discuss further in the following chapter. To make reference to commoditization is to express concern about the mixture of persons and commodities—the hybrids—and thereby to express longing for a clear distinction, just like pro-market proponents try to guard the borders and esteemed mechanisms of their society through classificatory exercises. It is not clear, however, and it cannot be settled, once and for all, what is and what is not part of "a body" and how bodies relate to persons.[93] "Therefore, I contend that, rather than finding answers to such questions, we should acknowledge how the gift/market debate manifests a society trying to establish such "foundational categories" and borders. From the perspective of critical legal studies, Alan Hyde suggests that we "define bodies in the first place only when we are conflicted, as a society and often within ourselves. When body boundaries become problematic."[94] To understand the values at stake at these borders, we need to understand the historical rise of particular forms of problematization of the body and its potential for exchange. We need approaches less concerned with clearing up the mess and more concerned with understanding what is going on as the mess unfolds.[95] When propagating this approach, I have by implication prioritized the mechanisms, concerns, and values at stake when ubjects are exchanged at the expense of settling their moral status. The rest of this book is concerned with exploring this approach, but first I briefly outline how the gift-market distinction has already been criticized in the literature by other scholars.

Neither Gift Nor Market

From a legal and philosophical perspective, Harris has argued that since the interests and dynamics at stake in ubject exchange are so complex, a market model poorly addresses them.[96] The market framework is simply an inadequate analytical framework to understand what is at stake when people exchange "parts of their body." Harris finds a sentence like "I own my body" quite harmless—but of limited analytical value. Any child knows the difference between expressions such as "my ball" and "my teacher," he says, and when we extend the property idiom to "my liver," we should be able to see the difference as well. Exchange of organs and other ubjects involves much more complex sets of interests than exchange of tennis balls. Along the same lines, Dickenson finds it naïve to believe that property rights to one's body would make people follow some sort of market forces. As a consequence, several scholars have developed alternative analytical approaches.

Based on fieldwork in a brain bank, geographer Bronwyn Parry has shown how the distinction between gift economy and market economy is inadequate for describing the actual mixtures of exchange forms that she studied, just as sociologist Rene Almeling has made this point ethnographically in her studies of gamete exchange in the USA.[97] Thereby she places herself in continuation of what is in fact a long tradition in anthropology for describing the gift-commodity distinction as a false dichotomy.[98] I will describe this tradition in more detail in Chap. 2. In fact no object, and no ubject I will add, is in a stabile position as either tradable or nontradable: commoditization and inalienation are processes, and ubjects may hold an infinite number of positions. This point is succinctly made in a recent report from the British Nuffield Council based on a working party chaired by Marilyn Strathern.[99] In this report, it is suggested to think of policy solutions in terms of corresponding steps on an "intervention ladder" where financial means gradually become used as incentives but where they need not always work in this manner. I would add, however, that ubjects are not even confined to merely scale back and forth between gift and commodity states (a scale which frames value and meaning according to market thinking):[100] they operate also on a variety of other scales and can be sacred or profane; symbols of power, sexuality, or commemoration; or viewed as life-saving resources, means for knowledge production, or causes of infection.[101] Furthermore, the images of gifting employed in the arguments presented above are modeled on specific (predominantly Western) conceptions of gift, but "to give" can have many meanings and not least the giving of ubjects holds many layers of significance which are all under cultural and individual variation.[102] In his seminal essay *The Gift: The Form and Reason for Exchange in Archaic Societies*, Marcel Mauss argued that gifting is a basic institution involving a threefold obligation: to give, to receive, and to return. Exchange therefore involves more than material redistribution; it revolves around construction of relations. Though often read in the Anglophone world as an essay about exchange in an economic sense, Mauss primarily saw his essay as contributing to the study of primitive law delivering basic insight into the making of enduring relations.[103] With Mauss we might thus say that ubject exchange reflects more than economic value: it produces relations and interacts with institutional change. To capture how gamete exchange and fertility treatment involve so much more than economic value, Charis Thompson suggests thinking about the fertility field's intricate structuring of bodies, gametes, and kinship as an ontological choreography shaping both structure and agent.[104] Her insights are important because indeed many ubjects, as we shall see, can be seen as relation builders participating in an ontological choreography. Thompson illustrates why a framing of the type "markets in human body parts" misses essential elements of the exchange, and she is important as a central figure in a stream of ethnography that seeks to find alternatives to the market framing. However, it is important not to uplift the logic of relation building (found in Mauss and parts of the recent ethnographic tradition) to a universal principle. Many ubjects leave bodies without attracting attention or creating any relations or new social structures. They leave the body and fall more or less unnoticed into plain materiality. Furthermore, even ethnographers who explicitly argue that a market framing delivers an inadequate representation of the values involved in ubject exchange often draw upon implicit moral assessments typical of the antimarket position described above as a sort of commoditization hypothesis.

The distinction between two models, market and gift, is very much an effect of the immense influence of Titmuss' thinking.[105] Waldby and Mitchell find that Titmuss is outdated for five reasons:[106] (1) Today, it is rarely a stable substance like blood which is transferred but technologically mediated products. This distributes the agency going into making the "donation" and the "product."[107] (2) Most transfers are taking place at an international scale rather than as nationally controlled systems of procurement. (3) Trade is ubiquitous and not something a nation state can decide to ban. (4) Gifts and commodities are not opposites but presuppose each other. (5) Today, flows of bioinformation have taken over the scene from the substances that preoccupied Titmuss. I agree with Waldby and Mitchell on the need to move beyond the gift/market distinction and to see instead how traits associated

Ubjects include hip prothesis

with each regime are enfolded into each other in specific forms of exchange. But not all ubject exchange is destined on technological mediation and bio-informatization as they suggest. The examples they employ reflect a high-tech prejudice and a particular understanding of what is part of a body reflected also in their analytical category "*tissue economy.*" With this book and the category of the ubject which includes metal remains and mundane substance-oriented transplant ubjects like bone, I wish to bring other nuances to our contemplation of what ubject exchange sets in motion.[108] And I wish to make ethnographers who claim to have moved beyond the gift/market dichotomy when describing how "donated body parts turn into commodities" reconsider whether they have actually succeeded—or if they continue to aim for a "gift model" and to formulate moral complaints shaped by a historical ideal of total separation between persons and commodities. As I engage the intricacies of ubject exchange, I highlight the way in which it interacts with and changes ideas about bodies and persons as well as markets. *We are exchanging bodies*—providing change through exchange—and in this process neither the body nor the mode of exchange remains stable and untouched.[109]

Structure of the Book

Chapter 2 outlines the historical rise of market thinking and the ideas about persons and commodities that emerged from these processes. I use this outline to explain how ubjects challenge and provoke basic distinctions in a pervasive social institution. By contrasting market thinking with anthropological exchange theory, I illustrate further the specificity of market thinking and, thereby, some of the underlying reasons for current ubject exchange to cause so many concerns.

In Chap. 3, I suggest that if we wish to understand the social implications of current forms of ubject exchange, we need to develop ways of talking about the body that do not presume fixed boundaries. I draw on recent studies in STS, anthropology, and the health sciences to unfold the complexity of what ubject exchange involves and indicate how we might begin to theorize the ubject. The purpose of this chapter is on the one hand to move beyond naïve assumptions about the body proper, which are characteristic of some antimarket arguments of dignity, and on the other hand to show how ubject exchange involves issues that pro-market arguments rarely acknowledge. I am basically trying to develop an understanding of the relationship between ubjects and subjecthood and, thereby, what it is that make ubjects move in special ways. To claim—as some market proponents do—that ubjects are things, and things are material objects of plain monetary worth, and that all other views are superstition in itself represents a gross superstition of the type Latour talks about in *We Have Never Been Modern.* The chapter builds its argument on a synthesis of two strains of literature that have treated bodies quite differently, namely, phenomenology and poststructuralism, and makes an audacious (outright foolhardily) attempt of effecting this synthesis by drawing on their common roots in the German philosophers Arthur Schopenhauer and Friedrich Nietzsche, in particular their work on will.

I suggest that it is from people's experiences with willful beings that ubjects gain their potential for subjecthood. In conclusion, it is argued that how bodies are done reflects individual and societal attempts of self-delineation. We need to acknowledge the complexity of this process to avoid falling into some of the common pitfalls of the debate between market proponents and opponents.

Chapter 4 illustrates the intricate ways in which money is passed on and how the involved mechanisms are poorly appreciated if we subscribe to market thinking. The inclination to separate commodities and persons stimulates interesting attempts to make money transfers appear as something else than payment for the actual ubject. Through case studies, this chapter describes the process of price setting for that which is not supposed to be owned and sold. The scope of the argument is broadened by situating the findings from the case studies in a personal reading of the literature on public attitudes to trade of ubjects and relatively well-known narratives of ubject exchange. To finish, it is argued that we need to pay more attention to the moral work at stake in all the various attempts of making ubject exchange special and "noncommercial" if we wish to build better policies. It is not enough to claim that such attempts constitute market failures.

In Chap. 5, I combine the analytical insights from Chaps. 2 and 3 with empirical insights from Chap. 4 in a general discussion of what makes some ubjects attract attention at certain times and in particular situations, while others can pass in and out of bodies relatively unnoticed. Using the overall biopolitical framework, the chapter argues that when exchange in certain secular societies takes on a form that is seen as "market-like," it stimulates a potential for subjecthood, which can be seen as resting in all ubjects. Such transactions seem to *make* ubjects morally significant (rather than simply undermining their moral worth) because the underlying logic of market thinking enacts the dichotomy between persons and things. I suggest that many of the controversies surrounding ubjects illustrate attempts of making a distinction between worthy ends and plain extractable resources.

Finally, the Conclusion summarizes the overall argument and addresses the so-what question in relation to where this takes us vis-à-vis the pro and con debate about "markets in human body parts." Exchange creates change—both in bodies and in institutions—and these changes are important. The morality of market thinking is not given, and we need to consider what the current biopolitical reconfiguration might imply for the fundamental institutional logics that drive, and are driven by, ubject exchange. From a new analytical position, it is possible not only to provide new types of answers to the questions raised in the debate about "markets in human body parts" but also to rethink the questions as such.

Throughout, I employ a mixture of genres and argumentative styles. In anthropology it has been common for quite some time to be committed to a particular genre in which arguments gain force through evocative prose, as also mentioned above. I like this genre, but I fear it is part of generating the problems I seek to address in this particular debate about "markets in human body parts." I therefore opt for a different argumentative style. As a consequence, the pages that follow might seem slightly strange and unfamiliar, at least for an ethnographer accustomed to look for the values associated with ethnographic density: literary elegance, a sense of presence and authenticity, detail and context, as well as a keen insistence on arguing only what can be argued from your own fieldwork. I adopt an argumentative style that sometimes follows the ethnographic tradition while at other times takes on a more formal, philosophical ring more akin to deductive reasoning. I am aware that some ethnographers will think that some of the qualities most dear to them are absent, or downplayed at least, and yet I adopt this style intentionally. Despite the general preference in ethnographic circles for evocative prose (so that we can *feel* the uncanniness—or horror—associated with "bodies sold as commodities"), I wish to cool down the ethnography with reasoning of a more philosophical, economic, and legal type. Concurrently, I wish to warm up and destabilize the philosophical, economic, and legal pro-market arguments with ethnographic examples. To achieve both, and thus move beyond the divide in the debate, I need to combine genres. I hope it will not deflect readers; it is designed instead to provoke and perhaps even inspire some ethnographers to engage with work in legal studies, economics, and ethics and appreciate how other disciplines sometimes dismantle what ethnographers take for granted. The provocations I aim for are not emotional or moral; they are analytical. They serve to arrive at new ways of framing questions relating to the financial, emotional, and material aspects of bodily exchange relations. I hope such transgressions will at some point (again) become more common and accepted in anthropological circles, at least if we really wish to question what people mean when they write about markets, when they write about bodies.

Endnotes

1. BBC News (2005, 2008), Bone (2008), Powell and Segal (2006); see also Cheney (2006) and Warren (2006).
2. *The New York Times* (Brick 2005).
3. For example, Caplan (2007) and Taylor (2005a).
4. Beck (2008) and *Medical News Today* (2009).
5. Irin (2009); Kale (2008); see also Scheper-Hughes (2000, 2001a, 2005).
6. See, for example, Goodwin (2006).
7. Organisation for Economic Co-operation and Development (OECD) (2001, 2005, 2011).
8. Waldby and Mitchell (2006).
9. Lock (2002); see also Haddow (2005) and Kent et al. (2006).
10. Squier (2004:2).
11. Waldby (2000:19); see also Waldby (2002, 2006). Sarah Franklin (2007) has noted similar semantic overlaps in relation to the notion of *stock*, and Dillon and Lobo-Guerrero (2009) have described how *species* refer to classification, biology, and money.
12. Eiss and Pedersen (2002:283).
13. Knowles (1999).
14. See Taylor (2005b); note also how Cecily Palmer refers to tissue samples as ambiguous: "Human and object, subject and thing" (Palmer 2009:15).
15. A short personal history of the concept: I first heard the word mentioned in a commentary that Stefan Helmreich made at the AAA Conference in San Francisco in 2008. Helmreich had picked it up from historian Hillel Schwartz and seen it being used at an art show to refer to "unique objects" (http://www.schmidtartcenter.com/previous-exhibitions.html, last accessed May 13, 2011). At the time I had long struggled with finding a proper term for the phenomena

that this book explores, and inspired by Leigh Star's work (with coauthors) on boundary objects (Bowker and Star 1999; Star 1989; Star and Griesemer 1989), I had tried out, for example, the term *human boundary object* (Hoeyer 2010b). However, implicitly this terminology seemed to insist on preexisting boundaries and entities, and it designated the ubject as an object. I have also considered using existing terms such as actant or cyborg (which are however both too broad), quasi-object/quasi-subject (which is not specific enough and either necessitates emphasizing one over the other or is too long), bio-objects (which overemphasize both biology and objecthood), or, as suggested by Bharadwaj, bio-crossings (which also place too much weight on biology), or abject in Butler's Kristeva-inspired sense (which however focuses on the morally repugnant which need not be the case with what I call ubjects). Ubject first seemed like a fun term to play around with, and then at some point, I realized that it did exactly what I needed it to do for me. I sincerely thank Helmreich for introducing it to me!

16. There is a long-standing debate in anthropology about emic and etic concepts. Should one take local (emic) words and develop an ethnowgraphic understanding of what they entail, thus building an ethnographically grounded theoretical understanding of a particular phenomenon, or should one seek to establish (etic) universal terms for comparisons across contexts? Recently the weight has been on the emic approach, but though I do not claim any universal relevance for my conceptual approach, I suggest that it can be helpful when studying your own intellectual landscape to define a vocabulary that distances itself from and defamiliarizes some of the assumptions underlying the usual parlor.

17. The anthropological literature has mostly opted for the concept of property relations rather than "ownership" in descriptions of the relationships between and among persons and things (Hann 1998; Strathern 1999). According to this tradition, we must see property relations as complex webs of relations, not just between a person and a commodity but among persons expressing their entitlements vis-à-vis each other and the resources surrounding them.

18. See, for example, Charo (2004) and Grubb (1998). Courts are involved in settling these issues at a very practical level (Fox and McHale 2001), of course, but the property status of body parts attracts many academic legal reflections too (Björkman and Hansson 2006; Bovenberg 2006; Grubb 1998; Hardcastle 2009; Laurie 2002; Skegg 1975).

19. Fox (2000). Others seem intrigued rather than annoyed by the anxiety generated by the blurring of things and persons (Hyde 1997). See also Alain Pottage (2004:5) who notes "that persons and things have multiple genealogies, and … their uses are too varied to be reduced to one single institutional architecture. Each form or transaction constitutes persons/things in its own way."

20. Rose (1994).

21. For some, entitlements denote something resting with persons in the same way as the notion of rights has been used. I wish to emphasize that entitlements cannot exist in a social vacuum; they are always social entitlements. Entitlements can include noncommercial entitlements such as a right to give informed consent and conventional commercial entitlements such as intellectual property rights (IPR)—both administrative entitlements to decide who may use, for example, cell lines in their research and entitlements to take custody of such material.

22. Bergström (2000).

23. When exploring morality, Signe Howell (1997) (see also Strathern 1997) has suggested focusing on moral reasoning as a continual process rather than looking for underlying, presumably stable values. It is important moreover to remember that people's moral values operate beyond the spoken word.

24. See discussion in Hoeyer (2008).

25. For an interesting and thoughtful introduction to biopolitics, see Lemke (2009).

26. See in particular *The History of Sexuality* (Foucault 1986, 1992, 1994). The development in governmental logics is succinctly described in a lecture on governmentality (Foucault 1991), and the mechanisms involved in shaping spaces for action with different implications for different people enrolled into medical settings laid out with gruesome clarity in (especially chapter 5 of) *The Birth of the Clinic* (Foucault 1973).

27. Anthropologist David Graeber has encircled three streams of thought converging in the current uses of the word value: what is proper [ethics], what is desired so that people want to do something to get it [power], and what is meaningful difference [truth]—and I have in brackets indicated how they relate to the Foucaultian axes (Graeber 2001).

28. In fact I have published work in many different traditions. I think it is important to play around with genres and disciplinary traditions and for each topic chose how far you can go if you wish to reach a particular audience. In a sense, I think it radicalizes Foucaultian insights to avoid internalizing genre and custom as truths and instead operate in many different genres including those of a more positivist approach. I have conducted surveys and written more philosophical papers, all dependent on who I wanted to engage with a particular topic. Nevertheless, the willingness to write in different traditions (and sometimes use expressions that are slightly at odds with your normal thinking) reflects this ultimately fluid ontology which I do believe ought to stimulate multiple forms of inquiry rather than buttress a particular genre.

29. Latour (1986).

30. The Foucaultian study of power relations and perceptions of body and self comprises a huge corpus of literature, which it would take a book of its own to summarize (cf. Lemke 2009). A little strain of this literature involves the concrete politics and policies of bodily donations, which is an aspect I have consciously downplayed to allow more focus on the performativity of the moral agency feeding into such policies.

31. A fascinating study of the interplay of EU and British legislation which explores a number of the issues related to the debate at the center of this book can be found in Julie Kent's book on the regulation of regenerative medicine, *Regenerating Bodies* (Kent 2012).

32. See, for example, Cooper (2008) and Rajan (2006); see also discussion in Rose (2007) and Franklin (2003, 2007). Mitchell and Waldby (2010) argue that neglect of economic aspects is related to a tendency for analysts to focus too narrowly on the state and issues of citizenship.

33. Rabinow (1992:170).

34. Bynum (1995:17). See also Bynum (1991). Kathrine Park suggests that the notion of strong material continuity identified by Bynum is a feature primarily of Northern European theology at the time, which is actually partly explaining why anatomical dissection developed in Italy and not, for example, England or the German cities (Park 1995).

35. Jackson (1996) and Merleau-Ponty (2002:504).

36. Mol (1998) and Mol and Law (2004).

37. Latour (1993, 2004) and Serres (1982).

38. Gottweis et al. (2008:271). The quote was couched by Ingrid Metzler. Gottweis' approach to life science governance is a general source of inspiration for the biopolitical approach taken here (Gottweis 1998).

39. Freud (1978:241).

40. See the interesting work compiled in Niewöhner and Scheffer (2010). Langstrup and Winthereik (2010) among others in that volume exemplify how you might construe objects of comparison that are not presumed stable or easily transferable.

41. Strathern (2004).

42. Metaphors of fluidity are widely used in the STS literature to capture ontological emergence. John Law has (with coauthors) suggested that such metaphors might be too dependent on notions of continuity, unlike a metaphor of fire which can jump, disappear, and reappear (Law and Mol 2001; Law and Singleton 2005). I like wetlands for their specific, slobbery, material connotations, however (besides, taking it as a place in which to move, the sneaking dangers of a wetland are slightly more attractive than the burn of a fireplace). Furthermore, it is attractive considering Doyle's use of the term *wetware* to characterize bodily material in contrast to informational data (Doyle 2003).

43. Foucault (1997:231–2).

44. Jackson (1998).

45. Nietzsche (2000:236).

46. Mason and Laurie (2001:715). On the term "organ markets," see, for example, Kaserman and Barnett (2002) and on "fertility market," see Spar (2006).
47. See, for example, Becker (2009), Cherry (2005), Eisendrath (1992), Krawiec (2009), Satel (2008b), Schlitt (2002), and Taylor (2005a).
48. Cherry (2005:23).
49. Hardcastle (2009).
50. Herring and Chau discuss this point in great depth (Herring and Chau 2007). See below.
51. Mason and Laurie (2001:722).
52. Cf. Laurie (2002) and Mason and Laurie (2001).
53. Goodwin (2006). Along similar lines Reichardt (2009) suggests that by prohibiting sale, the state exploits the people willing to donate for altruistic reasons.
54. Fabre (2006).
55. Similar arguments have been suggested by Friedlaender (2002) and Roff (2011).
56. Cherry (2005:152).
57. Goodwin (2006), Satel (2008a), and Taylor (2008).
58. Satel (2008b:1).
59. Taylor (2005a:1 and 3).
60. Kaserman and Barnett (2002:84 and 98).
61. Cherry (2005), Taylor (2005a), and Wilkinson (2003).
62. Satel (2008b).
63. Kaserman and Barnett (2002).
64. Epstein (2008).
65. Kaserman and Barnett (2002:5).
66. See, for example, Taylor (2005a). Taylor's argument is different from Cherry's above in that it does not focus on the moment of selling as central to autonomy but on how new sources of income provide consumer autonomy.
67. Huang et al. (2008:32).
68. Kaserman and Barnett (2002:20).
69. Kaserman and Barnett (2002:69).
70. Bovenberg (2006) and Hardcastle (2009). Some scholars also argue that you can propagate a property model without supporting a right to sale. Property rights need not be affiliated with market modes of allocation (see Chap. 2).
71. Kant (1997:157).
72. Ibid.
73. See, for example, Munzer (1994, 1993). Munzer is skeptical toward Kant in one respect, however. He thinks Kant might fall prey to what he calls the fallacy of division: what goes for the whole need not go for the parts. The source of dignity is the ability for reason, Munzer writes, and since individual organs are not capable of reason, they could in principle be treated like things. Still, he argues that selling part of your body would exhibit lacking self-respect.
74. Perley (1992).
75. Dickenson (2005, 2007) and Radin and Sunder (2005); see also Moniruzzaman (2012), Muraleedharan et al. (2006), and Scheper-Hughes (2000).
76. Dickenson picks her examples to support her case and neglect the widespread use of male bodies in war, the overuse of men in clinical trials, and the way in which male gametes are objectified and exchanged with very limited concern for donor health interests.
77. Titmuss' (1997) argument is often repeated; see, for example, Murray (1987).
78. Healy (2006).
79. The general shift is described in Copeman (2005). Recent policies in the EU as well as the USA advocate gift models without any dedicated attempts of controlling the calculation of recovery costs (Hoeyer 2010a).
80. Margaret Radin (1996), in particular, has theorized the concept of commoditization. Mostly, however, social scientists refer to commoditization without explicitly theorizing the concept as such.

81. Dickenson (2007:6).
82. Andrews and Nelkin (2001), Everett (2002), Klinenberg (2001), Linke (2005), Morgan (2002), Parry and Gere (2006), Ridgeway (2004), Rose (2001), Scheper-Hughes (2000), and Sharp (2000).
83. Holland (2001:263).
84. Wilkinson (2000); see also discussion in Wilkinson (2003).
85. Some exceptions prevail. Schlich (2007), for example, specifically delineates his concept of commodification to the kind of *gaze* on the human body described by Wilkinson.
86. The following quotes are from pages 2 and 4 in Sharp (2007).
87. Sharp (2007). See also Sharp (2000).
88. Ibid.:74.
89. Ibid.:49.
90. Scheper-Hughes (2000, 2001b); see also Dickenson above.
91. Rose (2001).
92. Cunningham (1998) and McAfee (2003).
93. Herring and Chau (2007). See also the anthropological discussion of this issue in Strathern and Lambek (1998).
94. Hyde (1997:11).
95. Harris (1996).
96. Harris (1996).
97. Almeling (2007, 2009) and Parry (2008).
98. Mauss (2000). See also Ferguson (1988) and Frow (1997).
99. Nuffield Council on Bioethics (2011).
100. Kopytoff (1986); see also Everett (2002) and Everett (2007).
101. Consider, for example, the differences revealed through work on embryos as research objects (Morgan 2002), rumors about exchange of shrunken heads or blood as elements in power struggles (Rubenstein 2007; Weiss 1998), penis theft as a theological topic (Smith 2002), and relics as representatives of ecclesiastical authority (Esmark 2002)—and consider how ubjects are sometimes desired while at other times disposed of and meant to go away.
102. Notice, for example, Bop Simpson's work on gifting of body parts in Sri Lanka (Simpson 2004).
103. Mauss (2000); for the shifting readings of Mauss' essay over time, see Sigaud (2002).
104. Thompson (2005).
105. Tutton (2004).
106. Waldby and Mitchell (2006:22–26).
107. See also Parry and Gere (2006).
108. Ubject exchange involves much more than monetary and dignitary interests. One thing I will not go into much detail with but which constitutes another good example has been pointed out by legal scholar Jane Kaye (2006): storage of tissue bits interacts with legal systems due to potential forensic uses. The protection of the law is at stake in a very different sense than that captured by the pro and con market arguments.
109. Parry (2004) similarly argues that property relations and market forms are changed in the process of finding ways to negotiate entitlements to DNA. See also Calvert (2008).

References

Almeling R (2007) Selling genes, selling gender: egg agencies, sperm banks, and the medical market in genetic material. Am Sociol Rev 72(3):319–340
Almeling R (2009) Gender and the value of bodily goods: commodification in egg and sperm donation. Law Contemp Probl 72:37–58

Andrews L, Nelkin D (2001) Body bazaar: the market for human tissue in the biotechnology age. Crown Publications, New York

BBC News (2005) Alistair Cooke's bones "stolen". BBC News (online)

BBC News (2008) Plea deal in US body parts case. BBC News (online)

Beck M (2008) Ova time: women line up to donate eggs—for money. The Wall Street Journal

Becker G (2009) Allowing sale of organs will increase the number of donations. In: Egendorf LK (ed) Organ donation: opposing viewpoints. Gale Cengage Learning, Detroit, pp 61–67

Bergström L (2000) Who owns our genes? The concept of ownership. Nordic Committee on Bioethics, Copenhagen, pp 101–110

Björkman B, Hansson SO (2006) Bodily rights and property rights. J Med Ethics 32:209–214

Bone J (2008) Alistair Cooke bodysnatch ring head gets up to 54 years in jail. Times online, 27 June 2008

Bovenberg JA (2006) Property rights in blood, genes and data: naturally yours? Martinus Nijhoff Publishers, Leiden

Bowker GC, Star SL (1999) Sorting things out—classification and its consequenses. The MIT Press, Cambridge

Brick M (2005) Alistair cooke's bone were stolen for implementation, his family says. The New York Times, 23 Dec 2005 (online)

Bynum CW (1991) Fragmentation and Redemption: essays on gender and the human body in Medieval Religion. Zone Books, New York

Bynum CW (1995) The resurrection of the body in Western Christianity, 200-1336. Columbia University Press, New York

Calvert J (2008) The commodification of emergence: systems biology, synthetic biology and intellectual property. BioSocieties 3:383–398

Caplan AL (2007) Do no harm: the case against organ sales from living persons. In: Tan HP, Marcos A, Shapiro R (eds) Living donor transplantation. Informa Healthcare, New York, pp 431–434

Charo RA (2004) Legal characterizations of human tissue. In: Youngner SJ, Anderson MW, Schapiro R (eds) Transplanting human tissue: ethics, policy and practice. Oxford University Press, Oxford, pp 101–119

Cheney A (2006) Body brokers: inside America's underground trade in human remains. Broadway Books, New York

Cherry MJ (2005) Kidney for sale by owner: human organs, transplantation, and the market. Georgetown University Press, Washington, DC

Cooper M (2008) Life as surplus: biotechnology and capitalism in the neoliberal era. University of Washington Press, Seattle

Copeman J (2005) Veinglory: exploring processes of blood transfer between persons. J R Anthropol Insit 11:465–485

Cunningham H (1998) Colonial encounters in postcolonial contexts. Crit Anthropol 18(2):205–233

Dickenson D (2005) Human tissue and global ethics. Genomics Soc Policy 1(1):41–53

Dickenson D (2007) Property in the body: feminist perspectives. Cambridge University Press, New York

Dillon M, Lobo-Guerrero L (2009) The biopolitical imaginary of species-being. Theory Cult Soc 26(1):1–23

Doyle R (2003) Wetwares: experiments in postvital living. University of Minnesota Press, Minneapolis

Eisendrath CR (1992) Used body parts: buy, sell or swap? Transplant Proc 24(5):2212–2214

Eiss PK, Pedersen D (2002) Introduction: values of value. Cult Anthropol 17(3):283–290

Epstein RA (2008) Altruism and valuble consideration in organ transplantation. In: Satel S (ed) When altruism isn't enough: the case for compensating kidney donors. The AEI Press, Washington, DC, pp 79–95

Esmark K (2002) De Hellige Døde og Den Sociale Orden: Relikviekult, Ritualisering og Symbolsk Magt. University Center of Roskilde, Roskilde

Everett M (2002) The social life of genes: privacy, property and the new genetics. Soc Sci Med 56:53–65

Everett M (2007) The "I" in the gene: divided property, fragmented personhood, and the making of a genetic privacy law. Am Ethnol 34(2):375–386

Fabre C (2006) Whose body is it anyway? Clarendon, Oxford

Ferguson J (1988) Cultural exchange: new developments in the anthropology of commodities. Cult Anthropol 3(4):488–513

Foucault M (1973) The birth of the clinic—an achaeology of medical perception. Vintage Books, New York

Foucault M (1986) The care of the self. Penguin, London

Foucault M (1991) Governmentality. In: Burchell G, Gordon C, Miller P (eds) The foucault effect: studies in governmentality. The University of Chicago Press, Chicago, pp 87–104

Foucault M (1992) The use of pleasure. Penguin, London

Foucault M (1994) Viljen til Viden. Seksualitetens Historie 1 [The will to knowledge: history of sexuality 1]. Det Lille Forlag, Copenhagen

Foucault M (1997) On the genealogy of ethics: an overview of work in progress. In: Rabinow P (ed) Ethics: essential works of foucault 1954–1984, vol 1. Penguin, London, pp 253–280

Fox M (2000) Pre-persons, commodities or cyborgs: the legal construction and representation of the embryo. Health Care Anal 8:171–188

Fox M, McHale J (2001) Regulating human body parts and products. Health Care Anal 8:83–86

Franklin S (2003) Ethical biocapital: new strategies of cell culture. In: Franklin S, Lock M (eds) Remaking life and death: toward and anthropology of the biosciences. School of American Research Press/James Currey, Santa Fe, pp 97–127

Franklin S (2007) Dolly mixtures: the remaking of genealogy. Duke University Press, Durham/London

Freud S (1978) The 'uncanny'. In: Strachey J (ed) The uncanny. Hogarth, London, pp 219–252

Friedlaender MM (2002) The right to sell or buy a kidney: are we failing our patients? Lancet 359(9310):971–973

Frow J (1997) Gift and commodity: time and commodity culture. Essays in cultural theory and postmodernity. Clarendon, Oxford, pp 102–217

Goodwin M (2006) Black markets: the supply and demand of body parts. Cambridge University Press, New York

Gottweis H (1998) Governing molecules. the discursive politics of genetic engineering in Europe and the United States. The MIT Press, Cambridge, MA

Gottweis H, Braun K, Haila Y, Hajer M, Loeber A, Metzler I, Reynolds L, Schultz S, Szerszynski B (2008) Participation and the new governance of life. BioSocieties 3:265–286

Graeber D (2001) Toward an anthropological theory of value: the false coin of our own dreams. Palgrave, New York

Grubb A (1998) 'I, me, mine': bodies, parts and property. Med Law Int 3:299–317

Haddow G (2005) The phenomenology of death, embodiment and organ transplantation. Sociol Health Illn 27(1):92–113

Hann CM (1998) Introduction: the embeddedness of property. In: Hann CM (ed) Property relations: renewing the anthropological tradition. Cambridge University Press, Cambridge, pp 1–47

Hardcastle R (2009) Law and the human body: property rights, ownership and control. Hart Publishing, Oxford/Portland

Harris JT (1996) Who owns my body. Oxf J Leg Stud 16(1):55–84

Healy K (2006) Last best gift: altruism and the market for human blood and organs. The University of Chicago Press, Chicago

Herring J, Chau P-L (2007) My body, your body, our bodies. Med Law Rev 15:34–61

Hoeyer K (2008) What is theory, and how does theory relate to method? In: Vallgårda S, Koch L (eds) Research methods in public health. Gyldendal Akademisk, Copenhagen, pp 17–42

Hoeyer K (2010a) An anthropological analysis of European Union (EU) health governance as biopolitics: the case of EU tissues and cells directive. Soc Sci Med 70:1867–1873

Hoeyer K (2010b) Anthropologie des objets-frontières humains: explorer de nouveaux sites pour la négociation de l'identité. Sociologie et sociétés 42(2):67–89

Holland S (2001) Contested commodities at both ends of life: buying and selling gametes, embryos, and body tissues. Kennedy Inst Ethics J 11(3):263–284

Howell S (1997) Introduction. In: Howell S (ed) The ethnography of moralities. Routledge, London

Huang E, Thakur N, Meltzer D (2008) The cost-effectiveness of renal transplantation. In: Satel S (ed) When altruism isn't enough: the case for compensating kidney donors. The AEI Press, Washington, DC, pp 19–33

Hyde A (1997) Bodies of law. Princeton University Press, Princeton

Irin (2009) Egypt: selling a kidney to survive. Irin, Cairo

Jackson M (1996) Introduction: phenomenology, radical empiricism, and anthropological critique. In: Jackson M (ed) Things as they are: new directions in phenomenological anthropology. Indiana University Press, Bloomington, pp 1–50

Jackson M (1998) Minima ethnographica: intersubjectivity and the anthropological project. University of Chicago Press, Chicago

Kale (2008) Poverty forces organ selling in Egypt. Poverty News Blog, 16 Dec 2008

Kant I (1997) Lectures on ethics: Immanuel Kant. Cambridge University Press, Cambridge

Kaserman DL, Barnett AH (2002) The U.S. organ procurement system: a prescription for reform. The AEI Press, Washington, DC

Kaye J (2006) Police collection and access to DNA samples. Genomics, Soc Policy 2(1):16–27

Kent J (2012) Regenerating Bodies. Tissue and cell therapies in the twenty-first century. Routledge, New York

Kent J, Faulkner A, Geesink I, Fitzpatrick D (2006) Culturing cells, reproducing and regulating the self. Body Soc 12(2):1–23

Klinenberg E (2001) Bodies that don't matter: death and dereliction in Chicago. Body Soc 7(2–3):121–136

Knowles LP (1999) Property, progeny, and patents. Hastings Cent Rep 29(2):38–40

Kopytoff I (1986) The cultural biography of things: commoditization as process. In: Appadurai A (ed) The social life of things: commodities in cultural perspective. Cambridge University Press, Cambridge, pp 64–91

Krawiec KD (2009) Sunny samaritans and egomaniacs: price-fixing in the gamete market. Law Contemp Probl 72:59–90

Langstrup H, Winthereik BR (2010) Producing alternative objects of comparison in healthcare: following a web-based technology for asthma treatment though the lab and the clinic. In: Scheffer T, Niewöhner J (eds) Thick comparison: reviving the ethnographic aspiration. BRILL, Boston, pp 103–128

Latour B (1986) The powers of association. In: Law J (ed) Power, action and belief—a new sociology of knowledge? Routledge & Keagan Paul, London, pp 264–280

Latour B (1993) We have never been modern. Harvard University Press, Cambridge

Latour B (2004) Gabriel Tarde og det Sociales Endeligt. Distinktion 9:33–47

Laurie G (2002) Genetic privacy: a challenge to medico-legal norms. Cambridge University Press, Cambridge

Law J, Mol A (2001) Situating technoscience: an inquiry into spatialities. Environ Plann D Soc Spaces 19:609–621

Law J, Singleton V (2005) Object lessons. Organization 12:331–355

Lemke T (2009) Biopolitik: En introduktion. Hans Reitzels Forlag, København

Linke U (2005) Touching the corpse: the unmaking of memory in the body museum. Anthropol Today 21(5):13–19

Lock M (2002) Human body parts as therapeutic tools: contradictory discourses and transformed subjectivities. Qual Health Res 12(10):1406–1418

Mason JK, Laurie GT (2001) Consent or property? Dealing with the body and its parts in the shadow of Bristol and Alder Hey. Mod Law Rev 64(5):710–729

Mauss M (2000) The gift: the form and reason for exchange in archaic societies. Routledge, London

McAfee K (2003) Neoliberalism on the molecular scale: economies and genetic reductionism in biotechnology battles. Geoforum 34:203–219

Medical News Today (2009) Clinics report rise in egg, sperm donations during recession. Medical News Today (online)

Merleau-Ponty M (2002) Phenomenology of perception. Routledge, London

Mitchell R, Waldby C (2010) National biobanks: clinical labor, risk production, and the creation of biovalue. Sci Technol Hum Values 35(3):330–355

Mol A (1998) Missing links, making links: the performance of some atheroscleroses. In: Berg M, Mol A (eds) Differences in medicine: unraveling practices, techniques, and bodies. Duke University Press, Durham/London, pp 144–165

Mol A, Law J (2004) Embodied action, enacted bodies: the example of hypoglycaemia. Body Soc 10(2/3):43–62

Moniruzzaman M (2012) "Living cadavers" in Bangladesh. Med Anthropol Q 26(1):69–91

Morgan L (2002) "Properly disposed of": a history of embroyo disposal and the changing claims on fetal remains. Med Anthropol 21(3):247–274

Munzer SR (1993) Kant and property rights in body parts. Can J Law Jurisprud 6(2):319–341

Munzer SR (1994) An uneasy case against property rights in body parts. Soc Philos Policy 11(2):259–286

Muraleedharan VR, Jan S, Prasad SR (2006) The trade in human organs in Tamil Nadu: the anatomy of regulatory failure. Health Econ Policy Law 1:41–57

Murray TH (1987) The gifts of the body and the needs of strangers. Hastings Center Rep 17(April):30–38

Nietzsche F (2000) Beyond good and evil: prelude to a philosophy of the future. In: Kaufmann W (ed) Basic writings of Nietzsche. Random House, New York, pp 179–435

Niewöhner J, Scheffer T (2010) Thickening comparison: on the multiple facets of comparability. In: Scheffer T, Niewöhner J (eds) Think comparison: reviving the ethnographic aspiration. BRILL, Boston

Nuffield Council on Bioethics (2011) Human bodies: donation for medicine and research. Nuffield Council on Bioethics, London, pp 1–254

Organisation for Economic Co-operation and Development (OECD) (2001) Biological resource centres: underpining the future of life sciences and biotechnology. OECD Directorate for Science, Technology and Industry, Paris

Organisation for Economic Co-operation and Development (OECD) (2005) Proposal for a major project on THE BIOECONOMY IN 2030: a policy agenda. OECD, Paris

Organisation for Economic Co-operation and Development (OECD) (2011) The Bioeconomy to 2030: designing a policy aganda. OECD, Paris

Palmer C (2009) Human and object, subject and thing: the troublesome nature of human biological material (HBM). In: Bauer S, Wahlberg A (eds) Contested categories, life sciences in society. Routledge, London, pp 15–30

Park K (1995) The life of the corpse: division and dissection in late medieval Europe. J Hist Med Allied Sci 50(1):111–132

Parry B (2004) Bodily transactions: regulating a new space of flows in "bio-information". In: Verdery K, Humphrey C (eds) Property in question: value transformation in the global economy. Berg, Oxford, pp 29–68

Parry B (2008) Entangled exchange: reconceptualising the characterisation and practice of bodily commodification. Geoforum 39:1133–1144

Parry B, Gere C (2006) Contested bodies: property models and the commodification of human biological artefacts. Sci Cult 15(2):139–158

Perley SN (1992) From control over one's body to control over one's body parts: extending the doctrine of informed consent. N Y Univ Law Rev 67(2):335–365

Pottage A (2004) Introduction: the fabrication of persons and things. In: Pottage A, Mundy M (eds) Law, anthropology, and the constitution of the social: making persons and things. Cambridge University Press, Cambridge, pp 1–39

Powell M, Segal D (2006) In New York, a grisly traffic in body parts: illegal sales worry dead's kin, tissue recipients. Washington Post (online)

Rabinow P (1992) Severing the ties: fragmetation and dignity in late modernity. Knowl Soc
    Anthropol Sci Technol 9:169–187
Radin MJ (1996) Contested commodities. Harvard University Press, Cambridge
Radin MJ, Sunder M (2005) The subject and object of commodification. In: Ertman MM, Williams
    JC (eds) Rethinking commodification: cases and readings in law and culture. New York
    University Press, New York, pp 8–29
Rajan KS (2006) Introduction: capitalisms and biotechnologies. Biocapital: the constitution of
    postgenomic life. Duke University Press, London, pp 1–36
Ridgeway J (2004) It's all for sale—the control of global resources. Duke University Press,
    Durham/London
Reichardt JO (2009) Donor compensation. An ethical imperative. Transplant proc 42:124–125
Roff RS (2011) We should consider paying kidney donors. BMJ 343:d4867
Rose CM (1994a) Introduction: approaching property. In: Property and persuasion: essays on the
    history, theory and rhetoric of ownership. Westview Press, Boulder/Oxford, pp 1–8
Rose H (2001) The commodification of bioinformation: the Icelandic health sector database. The
    Wellcome Trust, London
Rose N (2007) The politics of life itself: biomedicine, power, and subjectivity in the twenty-first
    century. Princeton University Press, Princeton
Rubenstein SL (2007) Circulation, accumulation, and the power of shuar shrunken heads. Cult
    Anthropol 22(3):357–399
Satel S (2008a) Concerns about human dignity and commodification. In: Satel S (ed) When altru-
    ism isn't enough: the case for compensating kidney donors. The AEI Press, Washington, DC,
    pp 63–78
Satel S (2008b) Introduction. In: Satel S (ed) When altruism isn't enough: the case for compensat-
    ing kidney donors. The AEI Press, Washington, DC, pp 1–10
Scheper-Hughes N (2000) The global traffic in human organs. Curr Anthropol 41(2):191–224
Scheper-Hughes N (2001a) Bodies for sale—whole or in parts. Body Soc 7(2–3):1–8
Scheper-Hughes N (2001b) Commodity fetishism in organ trafficking. Body Soc 7(2–3):31–62
Scheper-Hughes N (2005) The last commodity: post-human ethics and the global traffic in "fresh"
    organs. In: Ong A, Collier SJ (eds) Global assemblages: technology, politics, and ethics as
    anthropological problems. Blackwell Publishing, Oxford, pp 145–168
Schlich T (2007) The technological fix and the modern body: surgery as a paradigmatic case. In:
    Crozier I (ed) The cultural history of the human body, vol 6 "1920–present". Bergh Publishers,
    London
Schlitt HJ (2002) Paid non-related living organ donation: horn of plenty or Pandora's box? Lancet
    359(9310):906–907
Serres M (1982) The parasite. John Hopkins University Press, Baltimore
Sharp LA (2000) The commodification of the body and its parts. Annu Rev Anthropol
    29:287–328
Sharp LA (2007) Bodies, commodities, and biotechnologies: death, mourning, and scientific desire
    in the realm of human organ transfer. Columbia University Press, New York
Sigaud L (2002) The vicissitudes of *The Gift*. Soc Anthropol 10(3):335–358
Simpson B (2004) Impossible gifts: bodies, Buddhism and bioethics in contemporary Sri Lanka.
    R Anthropol Inst 10(4):839–859
Skegg PDG (1975) Human corpses, medical specimens and the law of property. Anglo-Am Law
    Rev 4(4):412–424
Smith M (2002) The flying phallus and the laughing inquisitor: penis theft in the Malleus
    Maleficarum. J Folk Res 39(1):85–117
Spar DL (2006) The baby business: how money, science, and politics drive the commerce of
    conception. Harvard Business School Press, Boston
Squier SM (2004) Liminal lives. Duke University Press, Durham

Star SL (1989) The structure of ill-structured solutions: boundary objects and heterogeneous distributed problem solving. In: Gasser L, Huhns Michael N (eds) Distributed artificial intelligence, vol II. Morgan Kaufmann Publishers, Inc, San Mateo, pp 37–54

Star SL, Griesemer JR (1989) Institutional econology, "translations" and boundary objects: amateurs and professionals in Berkeley' museum of vertebrate zoology, 1907–39. Soc Stud Sci 19(3):387–420

Strathern M (1997) Double standards. In: Howell S (ed) The ethnography of moralities. Routledge, London

Strathern M (1999) Property, substance and effect: anthropological essays on persons and things. The Athlone Press, London

Strathern M (2004) Partial connections. AltaMira Press, Walnut Creek

Strathern A, Lambek M (1998) Introduction—embodying sociality: Africanist-Melanesianist comparisons. In: Lambek M, Strathern A (eds) Bodies and persons: comparative perspectives from Africa and Melanesia. Cambridge University Press, Cambridge, pp 1–25

Taylor JS (2005a) Stakes and kidneys: why markets in human body parts are morally imperative. Ashgate, Hampshire

Taylor JS (2005b) Surfacing the body interior. Annu Rev Anthropol 34:741–756

Taylor JS (2008) Donor compensation without exploitation. In: Satel S (ed) When altruism isn't enough: the case for compensating kidney donors. The AEI Press, Washington, DC, pp 50–62

Thompson C (2005) Making parents: the ontological choreography of reproductive technologies. The MIT Press, Cambridge

Titmuss R (1997) The gift relationship: from human blood to social policy. The New Press, New York

Tutton R (2004) Person, property and gift: exploring languages of tissue donation to biomedical research. In: Tutton R, Corrigan O (eds) Genetic databases: socio-ethical issues in the collection and use of DNA. Routledge, London

Waldby C (2000) The visible human project: informatic bodies and posthuman medicine. Routledge, London

Waldby C (2002) Stem cells, tissue cultures and the production of biovalue. Health 6(3):305–323

Waldby C (2006) Umilical cord blood: from social gift to venture capital. BioSocieties 1:55–70

Waldby C, Mitchell R (2006) Tissue economies: blood, organs, and cell lines in late capitalism. Duke University Press, London

Warren J (2006) BTS stolen body parts scandal generating gruesome headlines, fears of infection; NY grand jury meeting. Transplant News (online)

Weiss B (1998) Electric vampires: Haya rumors of the commodified body. In: Lambek M, Strathern A (eds) Bodies and persons: comparative perspectives from Africa and Melanesia. Cambridge University Press, Cambridge, pp 172–194

Wilkinson S (2000) Commodification arguments for the legal prohibition of organ sale. Health Care Anal 8(2):189–201

Wilkinson S (2003) Bodies for sale: ethics and exploitation in the human body trade. Routledge, London

Chapter 2
What Is a Market?

This chapter outlines the tenets of market thinking to provide a better background for understanding the type of problem ubjects constitute and one of the institutions potentially changed through their exchange. Activities which are today usually described as "following the rules of the market" could have been conceived of in many ways, and to understand the specificity of what prevalent ideas about "markets" do to our way of handling ubjects, we need to elucidate the underpinnings and implicit assumptions associated with them. I pointed out in Chap. 1 how the concept of "market" has historically specific meanings. All concepts of course have historically specific meanings and as we reflect on them and as the practices they refer to change, these meanings also change.[1] The problem with market thinking is not that it is historically specific but that it implies particular distinctions that are at odds with ubject exchange. To better appreciate the implications of the resulting tensions, we need to *denaturalize* market thinking as a transparent representation of exchange while simultaneously taking into account how notions of markets are embedded in very *concrete* experiences and material exchange practices and have performative effects for them.

Two strains of literature deal quite differently with the relationship between "thinking" as consciousness and "exchange" as a material practice. Scholars inspired by the textual turn in historiography and social science often place emphasis on discourses and intellectual history. When following this approach, one will tend to see market thinking as determining exchange practices and attempt to trace the origination of specific problematizations to understand why exchanges take on specific forms in specific periods of time. To some extent, however, they overplay the role of cognition as a determinant of all the doing in the world. Conversely, a materialist tradition has followed Marx as he put emphasis on social history as a material practice and announced that life "is not determined by consciousness, but consciousness by life."[2] Here the material mode of production is a base determining intellectual life as superstructure. Important insights emanate from both strains of literature, but I subscribe wholeheartedly to neither. Rather, I wish to draw on both. I will highlight the interconnectedness of ideas and material practices based on the assumption that meaning and material practices are coproduced. When people refer to "markets,"

K. Hoeyer, *Exchanging Human Bodily Material: Rethinking Bodies and Markets*,
DOI 10.1007/978-94-007-5264-1_2, © Springer Science+Business Media Dordrecht 2013

they will have concrete experiences making the word meaningful, and when they sometimes get appalled by the same vocabulary used in conjunction with bodies, it informs us about what they expect of a "market transaction" and how they want bodies to be treated differently. It also modulates how they perform certain exchanges, even if the vocabulary they use does not fully determine exchange flows.

I first concentrate on the more discursive dimension and present a quick outline of what we might consider a Weberian ideal type of market thinking. I wish to suggest that market thinking embodies a dichotomy between persons and commodities and that with awareness of this dichotomy, which I will refer to as a moral paradigm, we are better equipped to understand the debate outlined in Chap. 1. Following the outline of market thinking, I present some relatively well-known theoretical alternatives to market thinking and historical examples which can help the reader to understand my preference for the notion of exchange as well as allow appreciation of the specificity of the market idiom. Based on this discussion, I then summarize the historical rise of this characteristic dichotomy between persons and commodities before I briefly contemplate when and how ubjects became problematized as raw materials in the circuits of exchange today known as markets.

The Moral Paradigm of Market Thinking

When today reference is made to "markets," a mode of thinking is often engaged which construes any kind of "trade" as a natural phenomenon abiding to "natural laws" or "market forces," as already illustrated in Chap. 1. The underlying ideas are in many ways constitutive for the organization of Western societies as argued by, among others, Michel Callon and James Carrier.[3] Carrier delivers an excellent starting point for an analysis of what is at stake in the current rise of ubject exchange in the so-called bioeconomy when he characterizes central components in market thinking. According to Carrier, it is possible to identify seven components comprising a cultural logic that he calls the ideology of the free market and which corresponds to what I prefer calling market thinking.

The first and perhaps most basic assumption of market thinking is that the world consists of free individuals. Second, and associated with this individualism, is the presumption that is it irrelevant why people want what they want, as long as they can afford what they want. Third, people are expected to act rationally, and this implies wanting more for less. Fourth, this leads to a negotiation of price in which buyers and sellers have conflicting interests. Fifth, this negotiation involves a dynamics of supply and demand reflecting the relative bargain power of the buyer and the seller. Sixth, choice—in particular consumer choice—is a moral good which implies also that more choices are better than fewer choices. Seventh, since sellers will try to fulfill demands, they will innovate and compete as well as increase their efficiency to win market shares. Market thinking thereby takes point of departure in desires located in people and is dedicated to the fulfillment of these desires. This view of desire as *causa sui* stands in contrast to a Foucauldian thinking which typically presumes a social shaping of desires. In the following chapter, I will suggest an alternative mediating understanding of desire through a conceptualization of will.

In market thinking, desire fulfillment is a moral good, and a free market—understood as a place where everybody is free to buy and sell what they want and can—is the best way of reaching this end.[4] In this sense, market thinking is a morally infused prescription for how societies ought to organize. A free market is expected to deliver a *fair* distribution of goods because people's assets reflect merits and efforts which in turn represent the work people do to fulfill desires. Equilibrium is something to be achieved through negotiations between individuals, not through governmental decree. Market thinking thereby prescribes a procedure for delivering an answer to otherwise morally contested topics such as "What is the right price?" or "What ought to be sold?" or "Who are entitled to receive which goods?"—topics of absolute urgency in much of the literature contemplating monetary incentives for organ donation. Questions about what is right or wrong, or what is need and what is fancy, can be replaced with a mechanism through which such questions are resolved. The resources people can use when negotiating price are related to work and property, and work can be converted to property and thus become accumulated with strengthened bargaining power as a result. Because property rights link people to goods and the basis of accumulation, they are the very cornerstone of "markets." Without them the other mechanisms could not function.

Various elements of these ideas feature in the pro-market arguments outlined in the introductory chapter: if obstacles for a free price setting (e.g., culture or ideas about dignity) are removed, people will be motivated to sell what they have and supply will increase, distribution be fairer, and by granting property rights to donors, a better position for bargaining and thereby reaching equilibrium between partners can be ensured. Market thinking is not a fixed set of propositions, however, and people may draw on some elements of it and still not subscribe to, for example, its moral evaluation. Like other discursive formations, market thinking provides paths through a moral landscape but cannot determine how they are used.

A Dichotomy Between Persons and Commodities

I wish to suggest that market thinking is morally infused in another manner too. It is premised on a dichotomy between persons and commodities. This dichotomy is not dealt with in any detail in Carrier's or Callon's work, but it is nevertheless essential to the legitimacy surrounding market thinking. Market thinking construes exchange as consisting of exchange partners and exchange objects linked by ownership. Ownership is in this understanding a relation between something owned and someone owning, usually one or several natural persons or corporations acting as legal persons. With this thinking, property comes to be viewed and understood through a capitalist lens of property relations, and thereby it acquires new meanings beyond the more restricted definition of commodities as industrially produced, identical objects manufactured for capitalist forms of exchange. Persons are the exchange partners and cannot be owned; commodities are the exchange objects.

Persons act; commodities (and any other form of tradable object) are acted on.[5] All legitimate objects of ownership come in on the commodity side of the equation whether they be raw materials, land, or animals. What counts as resources for commodity production depends on norms and available technologies, as pointed out by David Harvey, and changes over time.[6] What I am suggesting is that popular market thinking generates and upholds moral distinctions from an anthropocentric viewpoint. In this moral matrix, everything which is owned comes to represent a commodity, irrespective of the mode of production, and the commodity status is a way of signaling a status of "means" in a world where (wealthy) persons are ends. The ample use in the ethnographic literature of the word "commoditization" (also when clearly *not* dealing with industrially produced, identical objects manufactured for capitalist forms of exchange) underscores this point.

Respect for the entitlements of persons to property is a moral issue, and also in this sense a particular morality is constitutive for operational exchanges. When Kant first stated as moral dictum that "it is impossible to be a person and a thing, the proprietor and the property at the same time,"[7] he made a moral claim that spoke against existing slavery practices, but with the rise of market thinking as a pervasive cultural logic, this moral dichotomy has become a widely accepted tacit premise for the seven components identified by Carrier. It is so much taken for granted that it need not even be included into an analysis—we *know* persons are not to be treated as property. In market thinking, ownership is part of exerting undisputable entitlements over a given object. The role of state in ensuring and defining property rights is rarely articulated; ownership is presumed as a natural right. The owner of, for example, a bucket has unlimited rights to use, change, perhaps even destroy, or sell it as this person would see it fit. Actually, administrative law often restricts such rights, and from a legal perspective, ownership is actually a very open term: what it implies must be interpreted in light of its specific context. The basic point, which might seem banal, is that entitlements of the type exercised in relation to a bucket cannot be extended to persons without violating the basic structure of market thinking. Persons cannot legitimately be treated like commodities. Of course, we know that human beings still are enslaved around the world through various illegal exchange systems, but this type of exchange transgresses the ideal distinction between personhood and commodity, and it is deemed illegitimate by all major international organizations guarding human rights and free trade.[8] No NGOs are lobbying to gain official acceptance of slavery.

The ideal market can therefore be seen to consist basically of exchange partners (subjects) who have entitlements (ownership) to commodities (objects) circulating between them: a world of persons and a world of tradable things—as illustrated in Fig. 2.1.

This dichotomy corresponds to and gains strength from its association with other dichotomies. I have included some of them in the figure, and it will be obvious to the reader how they reflect also what Latour describes as the first dichotomy in the work of purification, the one between the social and the material. Absolute distinctions between the social and material are just illusions, and persons cannot be set aside from their materiality, as discussed in more detail in the next chapter, but here

Persons Subjects Active Culture	Linked by ownership	Commodities Objects Passive Nature

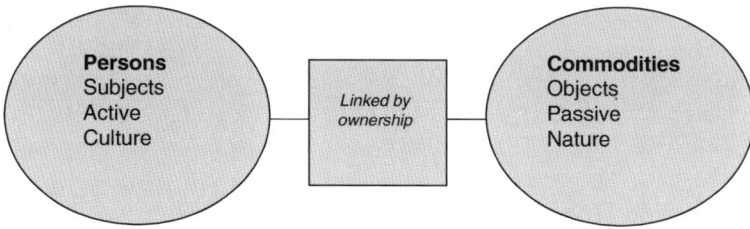

Fig. 2.1 The ideal distinction of the capitalist market

it is enough to state the obvious: as long as ubjects are seen by some as "part of persons," while also desired as "commodities," they transgress the moral paradigm characterizing market thinking.

For many types of exchange, this thinking is nevertheless powerful and useful. Market thinking *works*—in the sense that it delivers policy options and produces means of intervention. There is no such thing as a "market" existing out there on its own, following its own rules unmediated by how people themselves view the exchanges they take part in, but there are indeed a lot of exchanges taking a form that provides the concept of market with an almost phenomenological density. "Markets" are ingrained in accounting devices, juridical institutions, and physical infrastructures,[9] but though such devices might be scripted according to a particular set of ideas, they produce more than intended (as illustrated with the financial crises). Because it is so rigorously performed, "market" has become a meaningful concept not only for policymakers but also for people using it to make sense of their everyday experiences.

When ubjects are seen to challenge the moral paradigm underpinning market thinking, they also challenge the legitimacy of the exchange form. Unlike slave traders, however, the actors in the biotech industry actively pursue official recognition of their entitlements to ubjects (no further comparison intended). For example, some of the market proponents cited in Chap. 1 tried to define body parts as things; some research organizations lobby to extend patentability; some procurement organizations lobby to install financial incentives for ubject donation; and some processing companies lobby to secure their right to profit from the work they do when harvesting, processing, and distributing ubjects such as skin, tendons, and bone. Such attempts form part of continual developments of market thinking, and when they result in contestation, work on the moral legitimacy surrounding exchange is needed to sustain and stabilize the entitlements of the exchange partners. In fact, it need not be ubjects creating distortions to existing categories—we could just as well explain the intensified contestation of ambiguous bodily material as a result of purification work *creating* hybrids (ubjects) in the course of sustaining the power structures underpinning the market ideology. The moral status of a piece of bone was not important before a particular mode of production (here identified as market thinking) sought to establish a monopoly on categorization practices relating to value.

As I now turn to alternative ways of construing exchange, I seek to move beyond assumptions such as "the fact is that there *is* a market, and to say that there is not is to perpetuate a fiction"[10] while simultaneously getting closer to an understanding of the stakes in current disputes about ubject transgressions. Along the way, I will emphasize how market thinking is not devoid of morality—as some market opponents sometimes indicate—but rather relates to particular moral values currently undergoing change.

From Market Thinking to Analysis of Exchange

Because market thinking places such emphasis on price setting, it is typically accused for reducing exchange negotiations into a "how much" rather than a "what and how." Both Marx and Simmel saw money economy as a great equalizer of value through which all objects become exchangeable, or in the words of Oscar Wilde: "These days people seem to know the price of everything, and the value of nothing."[11] Still, the vision of a fully commercial society has never materialized. Anthropologists have studied the enduring cultural aspects of price-setting mechanisms even at ideal "market places" such as auction houses,[12] and economic sociologist Viviane Zelizer has been seminal in outlining social dimensions of "economic" activity and exposing problems with the "rigid concept of 'homo economicus' as a rusty, old-fashioned notion ready for retirement."[13]

A "market" is an ideal, not a name of a known phenomenon with identified borders and characteristics. Already Karl Marx showed that economies are not just out there ready to be discovered and described with the tools offered by classical economics; the notion of economy is historically specific and reflects a particular mode of production. Some scholars suggest that the idea of economy can be traced to the Anglo-American world in the seventeenth century.[14] Irrespective of whether it was really here economy first came to be construed as a separate field of human action, it did take centuries and a lot of intellectual labor to establish economy as a scholarly discipline in this part of the world.[15] In many ways, the emergence of something known as economy is interwoven with the academic discipline of economics, and "market" has gradually become an organizing term for much of this discipline. It is important, however, *not* to conflate market thinking with economics.[16] In fact, the sort of critique I am making is not directed at economics as such. Some of the ideas described in this chapter as "market thinking" have featured as elements of economic theories, of course, but while I avoid these theories as explanatory tools in my own account, I think it would be erroneous to dismiss the related economic theories altogether.[17] To dismiss economic theories because they do not deliver tools for understanding ubject exchange would be as foolish as dismissing anthropology or STS for not being able to estimate gross domestic products (GDPs).

Exchange Studies

The anthropological literature provides us with ample alternatives to market thinking to conceptualize the relations created through exchange. The *moral economy school* of the 1970s, for example, explained peasant conflict in Asia with social structures foreign to the idea of alienable property.[18] This placed institutions and social values above individual calculative agency in the understanding of changes produced by new types of exchange. The moral economy school was part of an academic debate between so-called *substantivists* who, inspired by Karl Polanyi, focused on the importance of values, institutions, and social structures and *formalists* who focused on the maneuvers of individual actors more in line with market thinking.[19] Substantivist anthropologists have described exchange systems that distinguish between different spheres of exchange in reflection of varying moral status of the exchanged objects.[20] Paul and Laura Bohannan, for example, described exchanges among the Tiv in West Africa as structured according to a three-layered multicentric economy: a subsistence sphere of food; a prestige sphere of cattle, slaves, expensive cloth, and brass rods; and a marital sphere for the exchange of women.[21] However, if in need, Tiv people might be forced to trade downward in moral status and exchange women for cattle or cattle for food items. Such transactions were viewed as morally degrading. In interesting ways this ethnographic work conceptualizes exchange in ways that appear useful also for understanding the type of conflict that ubjects instigate. We might consider, for example, the commoditization hypothesis described in Chap. 1 as an instance of normal moral reactions in a multicentric economy: ubjects are not beyond exchange, but they are supposed to flow in closed circuits of special moral value, and when they are traded downward, as they are when exchanged for money, it represents a moral transgression which is condemned.[22] As pointed out by Latour and Lépinay, however, substantivists in effect explain the mode of exchange with preexisting social structures whereby they underestimate the ways in which sociality is construed through differing modes of exchange. When exchanging ubjects, we not only use norms and institutions, we gradually change them.[23]

Formalists have challenged the concept of multicentric economies as overly focused on social structure.[24] In the 1980s, however, most anthropologists left the formalism/substantivism divide and began focusing instead on how objects travel back and forth between states of tradability and inalienability, between commoditization and singularization.[25] Other strains of anthropology have focused on ethnographic exploration of systems of circulation and accumulation to avoid presuming economy as a presupposed entity.[26] Maurice Bloch and Jonathan Parry argue that a recurrent feature in societies struggling to determine the legitimacy of various forms of exchange is how to balance egoism and community concern, how to ensure that circulation—in whatever form it takes—serves both individual and society.[27] They thereby point to exchange as a lens for understanding also the moral aspects of the emergence of social institutions. Beware of the differences between their notion of balance, however, and the assumptions about equilibrium in market thinking. Bloch and Parry do not relegate the involved agency to autonomous individuals with fixed amounts of monetary bargaining power and an assumption about the outcome being fair simply because it reflects negotiation.

Performative Effects

What this literature illustrates is that market thinking is not a transparent representation of how societies organize. We might, however, think of it as a cultural model *through which* they organize. How we organize as a society affects the ideas we use to make sense of society, and these ideas in turn have looping effects on our mode of organization.[28] With this framing, I follow Michel Callon who points to the performativity of market thinking for the phenomena it portrays.[29] Koray Çalışkan and Michel Callon have argued that substantivists and formalists alike bought into the notion of economy suggested by economists; only they disagreed about where to locate agency: in structures or agents?[30] Çalışkan and Callon suggest looking at processes of *economization* rather than presuming a phenomenon—economy—or a special type of activity designated by the adjective economic. Such processes consist of work and that work is performed by machines, institutions, and human actors alike. The work cannot be relocated to structure *or* agency nor can it be reduced to either meaning or materiality. Exchange is about movement and emergence and neither actor nor structure precedes the act of circulation; rather, actor and structure are results of incessant circulation of people and things (and what falls in between). We might use this to appreciate how ubjects are not just "holding economic value." Such value emerges through material processes of economization involving disentanglement, preparation, systems of distribution, and modes of consumptions.

Much of the anthropological debate about economics outlined above revolves around the problem of establishing an object of comparison—What can be compared if markets are not universal phenomena? Also historical analysis faces this problem, as pointed out also by Marx.[31] I do not claim universal validity of my alternative concepts of exchange, entitlements, and ubjects, but they are not quite as tainted in connection with the specific purpose of understanding what is at stake in current debates about something tagged as "markets in human body parts." I will now turn to a historical account illustrating the processes through which commodities as things, on the one hand, and persons together with their bodies, on the other hand, gradually became naturally opposing poles in an increasingly dominant moral paradigm which was not active in medieval exchanges.

The Historical Rise of the Person-Commodity Dichotomy

When, today, reference is made to commoditization as a repetitive element of history with medieval relics, anatomic dissections, and grave robbery as forerunners to the current surge in body products, it is wrongly asserted that relics at the time were conceived of as "commodities."[32] It has long been known that relics were

> quite literally part of the saint and still invested with his or her spiritual power. They were not mere symbols but active spiritual material, existing, as Christ had, in at least two places at once.[33]

Saints owned land, took part in conflicts, were physically exposed to violators of their entitlements, and expected to demonstrate their worth through their actions. Relics could be carried into infertile farmland, for example, and if their presence did not enhance the fertility of the spot, the power of the saint diminished.[34] By donating earthly wealth to the saint, the saint could be expected to intercede for the donors with the heavenly father. To enhance our understanding of this era of exchange, historical anthropologist Kim Esmark suggests greater appreciation of the way in which medieval exchange was intrinsically embedded in a specific cosmology. Drawing upon Mauss, Esmark analyzes the donation of earthly wealth to saints as total social facts, "at the same time juridical, economic, religious, and even aesthetic and morphological, etc."[35] Relics were not mere objects of exchange in a barter economy: the economy was organized according to the meanings ascribed to the relics and vice versa. Relics were spiritualized and subjectified and not just objectified, and even their monetary worth relied on this process of subjectification.

When the history of relics is retold as part of a genealogy of commoditization, it is informed by a conceptualization of commodities of more recent origin. The word commodity came into English in the fifteenth century derived from the French commodité. The concept was related to barter from the beginning, but as with the corresponding German concept, Ware, it acquired a special meaning in relation to the capitalist mode of production.[36] Today connotations to the word commodity are shaped by market thinking and experiences gained by interacting with the everyday practices of exchanges in accordance with this ideal. This thinking and these practices were foreign to the cosmology of medieval exchanges and depend upon a social and intellectual history that had not unfolded at the time.

Karl Marx's historical analysis of changing modes of production is a necessary point of passage when one seeks to understand the type of social change gradually taking place at the point in time when ideas about commodities emerged. Marx described the transcendence from a feudal mode of production to a capitalist one focusing on the industrialization in Great Britain beginning in the sixteenth century. A central component of Marx's analysis was the ways in which the capitalist mode of production alienated members of the working force from the fruits of their labor. In the capitalist mode of production, the commodity becomes a fetish,[37] gaining its own life independent of the hands making it. The commodity is "in the first place, an object outside us, a thing that by its properties satisfies human wants of some sort or another."[38] Marx limited the concept of commodities to objects produced with hired labor and the intention of trade. This is where he identified the alienation that facilitates *accumulation* of capital. New ways of accumulating capital (and thus power) are truly central to the changing mode of production. Labor is thus exchanged as part of a power game where the owner of capital is bound to win and gradually accumulate more profit.[39]

The changes in modes of production are interlinked with the rise of particular moral values, as also Max Weber argued in his seminal work on the spirit of capitalism.[40] Historian Thomas Haskell has argued that the type of trade associated with the capitalist mode of production also produced a particular type of experience of interacting with and depending on strangers and that these experiences in turn gave rise to support for a humanitarian morality.[41] Accordingly, Haskell argues, the emergence of a capitalist mode of production is connected to the abolition of slavery. Trade necessitated free workers: for capitalist exchange to become profitable, people had to be able to sell their labor. The stabilized legitimacy of the alienation of labor is thus intertwined with an emerging inalienability of the body. Besides, in a capitalist economy, people must serve not only as laborers but also as consumers, which is not the case in a slave-based economy. They need spare time for consumption. The new subject position involved a double function of the emerging working class as labor reserve and consumer outlet. My claim is that it was in the course of this historical process that *persons* as bodily agents and *commodities* as things became naturally opposing poles.[42]

Entitlements: From Authority to Property

Ideas about property rights—the salient link between persons and commodities in market thinking—also changed in this historical process. In the feudal system, entitlements to land were embedded in wider relations of mutual duties and obligations between sovereign, local lord, peasant, and peasant family members.[43] Scholars diverge in their views on whether the new understanding of property was a result of industrialization or a precondition for it,[44] but as already stated, I do not think one should try to make one of the two into a causal agent for the other. Still, it is a special feature of the capitalist mode of production to construe land and other goods as alienable commodities belonging to persons.[45] This alienation facilitated accumulation of private property in part of the bourgeoisie which thereby slowly acquired entitlements they could not have achieved in a system granting status and entitlements according to descent or theological authority.[46]

The complex developments in mode of production and property regimes evolved alongside intellectual transformations in notions of personhood and agency.[47] Foucault has described how "man" emerged as a new object of knowledge through the disciplines of biology, economy, and philology taking shape in the seventeenth and eighteenth centuries.[48] Concomitantly, the moral landscape of this "man" changed. Important intellectual and moral transformations in this period relate to the new importance attached to an individual's entitlements to the fruits of (predominantly) his or (not so often) her labor. These ideas are typically attributed to John Locke who commended work and aspiration and challenged the moral underpinnings of the feudal property rights. It has been argued that Locke in fact had very limited influence on his contemporaries: he was rarely cited in scholarly work at the time.[49] Still, gradually—with or without Locke's contribution—ideas about entitlements accruing from human agency were transfigured, and this had important consequences for property regimes and power relations. The later centrality of Locke indicates, perhaps, how posterity has attributed much greater moral importance to individual (human) contributions and looked for an early hero. A new subject position, a sort of autonomous agent, had gradually emerged: an individual dependent on personal ability rather than divine order.

Even if Locke did not hold such a central place among seventeenth-century intellectuals, they were indeed deeply engaged in the debates about the virtues associated with exchange, only along slightly different lines. John Pocock argues that "Economic man as a masculine conquering hero is a fantasy of nineteenth-century industrialization."[50] Well into the eighteenth century, the man focusing mainly on economic aspiration was seen as wrestling too much with his own passions and hysterias—and accordingly feminized. The mere commitment of one's labor to enhance one's wealth would not justify entitlements to its fruits in the same manner as market thinking today praises merit and personal efforts and thereby encourages additional work for the sake of fulfilling one's personal desires. However, since entitlements to property were gradually reconfigured in this period, so that profit from production became related to labor and capital rather than inherited entitlements, this emerging autonomous agent could begin to pursue (and produce) personal ambition in new ways. As capital was further detached from inheritance thanks to the system of public credit established by the Bank of England in the 1690s leading to what is known as the financial revolution in Great Britain (the credit system was in fact modeled on an older one established in 1609 in Amsterdam), it slowly became pertinent to construe autonomy not only as an intellectual and moral capacity but as a fair form of monetary self-generation reflecting only personal merits.

This image of the self-made (wo)man features strongly in public discourses today—perhaps stronger in the USA than elsewhere, but certainly with great legitimacy in many places.[51] The values attached to this self-reliance legitimize many actions along the lines of the argument "I should be allowed to do what I can *afford* to do," which also featured as the second element in Carrier's outline of market thinking. In this more recent form of market thinking, autonomy seems to be closely associated with consumer choice than the moral capacity for self-restraint originally argued by Kant in the eighteenth century ("I shall therefore not follow my inclinations, but bring them under a rule.").[52] Kant was much closely affiliated with a tradition emphasizing self-restraint, as described by Pocock. The contemporary legitimacy associated with pursuing whatever you can afford reflects continuing developments in ideas about autonomy into what we might call auto-referential agency: the active person is seen as making value out of a passively awaiting world.[53] Auto-referential agency is inherently anthropocentric: animals at a farm, for example, are not seen as laboring—they are managed; crops are not living by their own force—but grown. It has become coupled in interesting ways with capital ownership so that the one delivering capital is in effect "active" because decisions are seen as emanating from the one having the ultimate decision-making power (a logocentric notion of agency), whereas the one delivering only labor is not entitled to anything more than what he or she can achieve through negotiation.

Power, Morality, and Property Regimes

Throughout the period described here as the rise of industrialization, persons are not only construed in new ways as projects of knowledge and morality, they become different projects of power too. Up until the nineteenth century, most European states were under the rule of divinely sanctioned monarchs, but in the

USA and France (and gradually in the rest of Europe), state power became a secular force. Norbert Elias describes the rise of Western civilization as interlinked with a process of gradual creation of a state monopoly on physical power.[54] While the state assumes control of physical power, it provides another type of power: that of money. In case anybody does not respect the property rules that guide the use of money, the state is there to step in and supply physical power for its defense. State gradually begins to be thought of as facilitating conduct through provision of structured forms of freedom,[55] for example, the freedom to earn and spend money according to personal desires. The role of government became to conduct people's self-conduct—not to arrive at a divine teleological goal but to reach earthly progress. An early liberalism characterizing the secular state aimed at abolishing the privileges of the nobility and the guilds, while later liberalism (and in particular late-twentieth-century neoliberalism) came to be seen as the very antithesis of a state-centered society. These changes are widely debated in the governmentality literature. In his famous analysis of discipline and punishment, Foucault described a changing focus of power: from having centered on punishment of the body violating the rule of the sovereign, it came to center on disciplining the subject "behind" the act.[56] This new form of power rested on a Cartesian split between body and soul through which the will became a prime sphere of governmental intervention. The goal of government shifts from having been aimed primarily at sovereign control to revolving around facilitation of "population growth" and "economy."[57] The facilitating state was born and with it emerged the type of market thinking that, today, tends to portray *state* and *market* as opposing poles.

It seems to me that the rise of a market/state distinction converges with changing perceptions of legitimate spending. "Private" money has come to form part of a moral landscape of "autonomous" decisions aimed at hedonistic desire fulfillment because "public" money (and the "state") simultaneously took over some of the moral values and obligations relating to self-restraint and the common good.[58] With Bloch and Parry, we could perhaps think of this as a new balance between egoism and community concerns—but it is important not to mistake balance for justice or general welfare. I believe that this bifurcation can be helpful in understanding better some of the seemingly contradictory examples of current governmental decisions in the field of ubject regulation. When, for example, George Bush in August 2001 banned public funding of new embryonic stem cell lines, while allowing continued private funding, he was seen to take sides with the neoconservative pro-life movement in its defense of embryos as sacred. The apparent paradox in that he in fact gave up public control with embryonic stem cell research makes more sense if seen in light of the limiting of self-restraint to a public domain in combination with a deep entrenchment of ideas associated with the auto-referential agency (and its associated entitlements) in the private domain. He could act in self-restraint only on behalf of the "public" and make no such demands of the "private."[59] The industry "should be allowed to do what it can afford to do." Similar arguments reoccur in a number of decisions surrounding controversial uses of ubjects. In fertility treatment, for example, many countries have disapproved of public funding of controversial procedures while allowed private funding. A surprising number of seemingly leftish scholars also seem to subscribe to this logic when taking care to criticize only public spending on controversial technologies while apparently thinking that private spending is beyond reproach.

The paradoxical aspects of Bush's 2001 stem cell decision are not limited to loosening control while claiming to tighten it; he also made the embryos he supposedly wanted to use public powers to protect into natural resources for private, commercial exploitation.[60] This again must be understood against a backdrop of transfigurations already hinted at, namely, the disentanglement of "humans" (as agents) from "nature" (as materiality). For people adopting a Cartesian split between body and soul, bodies can apparently feature as part of an exploitable nature, but it is interesting that for Bush embryos were humans in the public domain and raw material in the private. It clearly illustrates the ambiguity of ubjects, but it is rarely appreciated in the ethics commentary. I take these paradoxes as instances of the productivity of the undefined and its role in changing institutions throughout a continuously developing social history.

There is no uniform and stable capitalist mode of production. The biotech industry, to which Bush wanted to transfer the responsibility of embryonic stem cells, is very differently construed from the type of industry Marx originally analyzed. Eugene Thacker, among others, has argued that a Marxist analysis of today's use of biological material in profit generation must take into consideration changes in capital, labor, and products.[61] Thacker suggests that even the notion of "industry" is misleading. Post-industrialization[62] was said to move emphasis from material production to information and services, but it was only a step on the way toward biotech industrialization in which capital is generated based on future prospects; labor consists of biomaterial labor performed by ambiguous biological entities and machines; and products involve what Thacker calls "life itself"[63] in informatized, optimized, and regenerated forms. The fusion of biology and capital involves

> …a set of complicated questions. For instance, in biomaterial labour, biology performs work, and yet there is no worker, no subject; biomaterial labour is strangely nonhuman. Who sells the labour power of biomaterial labour? We have a situation in which there is living labour without subjectivity.[64]

When Thacker uses the expression "strangely nonhuman," he indicates—implicitly—where the trouble resides. The food industry is all about making money out of biology, but this is rarely described as a "strange" transformation in the biocapitalism literature. For millennia, crops have performed labor on the fields and animals have done it in stables, but apparently, Thacker does not find it quite as conspicuous as the more recent biotechnological innovations. Hence, it is not biological labor as such, which is "strange." The ubjects typically doing part of the labor in the biotech industry are strange because they defy the usual categories of the moral paradigm of market thinking due to their partial connection to *persons*.[65] They thereby embody a potential which makes them relate differently to the capitalist structure than other types of material. As a consequence, this type of material typically enters the production through special means. It is not just appropriated through labor (as mining or farm products) but donated as gifts through informed consent procedures. And such gifts, which have once formed part of persons, are often said not to be owned: they are held in careful custody while worked upon. This *work* can be patented in accordance with the values associated with John Locke, and thereby the property holder comes to control the material and its use—without owning the material per se. Informed consent delivers a passage for the ubject from subjecthood to objecthood and does so by playing on the autonomous person that emerged with the capitalist mode of production. In relation to stem cell research, Melinda Cooper suggests the following:

> What is at stake here is a profound legal reconfiguration of the value of human biological life. The potential person will not be commodified—but the surplus life of the immortalized human stem cell will enter into the circuits of patentable invention.[66]

Elsewhere I have suggested that we might in fact rewrite the development of patent history as a long series of attempts of enacting the basic distinction between persons and commodities while allowing (human) biology to enter capitalist exchanges.[67] What we should notice here is not just that ubject exchange challenges the moral paradigm of market thinking but also the more basic point that "markets" are not static. Elements of market thinking might inform the structuring of the biotech mode of production by making particular types of entitlements appear morally legitimate and others illegitimate or uncanny—but there is not *one* capitalist mode of production, only perpetual reconfigurations of capital, labor, and products and, thereby, also of "markets," "persons," and "bodies."

Bodies: Hybrids and Purification

Some important questions remain unanswered: When and how did bodily material begin to enter the capitalist mode of production? When did ubjects become entangled in commercial processes of product making to the point where observers thought it reasonable to talk about a biotech mode of production or a bioeconomy in which biological components of human origin are constitutive for profit accumulation? These are questions that are easily treated far too monolithically in sweeping accounts as the one above. Thacker suggests that the biotech industry was born with the invention and patenting of genetic engineering by Genentech.[68] Complicating this dating is the much earlier commercial collection of blood, early tissue transplants, and procurement schemes for urine for hormone production, besides systematic procurement and distribution of breast milk. And of course there are the even earlier collections of specimens for research and teaching. A "complete history" of ubject exchange would require stable definitions over time, but what we might do is to trace shifting contestations of ubject entitlements over time in the juridical and other arenas.

Establishing a Legal Mandate for Ubject Exchange

There is not *one history* of *one body* in the legal system. It is simply impossible to pinpoint a particular event through which the body (re)entered the mode of production as raw material rather than labor.[69] Ubjects from dead bodies hold a different legal history than ubjects from living bodies; some groups of bodies have in some periods of time received protection that others have not; some ubjects have been viewed as part of persons by some but not others; and different people in different situations have attached varying hopes and fears to ubjects depending on the meanings attached to them and the opportunities available to the people involved. Rather than opting for a neat set of historical periods, one needs to understand the contingent, complex, strategic situations which in each case determine what can be exchanged by whom and how as well as who and what comes to serve as means for whom and what as ends. It is through complex and contradictory contestations that entitlements to bodies and their parts have taken shape. A history of problematizations necessarily revolves around public contestations, and it will not inform us about how bodies have been used behind the curtains, out of the limelight. Still, private acts are not irrelevant. They serve as an undercurrent of moral hybrids constantly traversing boundaries of public legitimacy, and through the contestations that they occasionally generate, new public moral positions emerge.[70]

Beginning with *living* bodies, it is generally said that slavery was abandoned in most European states during the medieval period. However, concomitant with the budding of the industrial revolution, the transatlantic slave trade gained enormous importance in international systems of exchange. Living bodies deemed *other*— Indians and Africans in the British Empire and Europeans in the Ottoman Empire— could be exchanged for money well into the nineteenth century. They were not regarded persons in the sense of the word tied up with market thinking. The abolition of slavery and its relationship to new understandings of personhood have been touched upon above, but it is remarkable that exchange of *dead* bodies for money seems to have become conspicuous according to a different set of logics and quite some time before abolition designated living (African and Indian) bodies as unfit for trade. In the medieval period, dead bodies were protected by ecclesiastical law rather than common law.[71] In 1644, in the UK, Coke in his writings termed the dead body *nullius in bonis* (no person's property), and this "no property rule" has been referred to regularly in courts in Great Britain from the eighteenth century onward.[72] Still, with the rise of anatomical dissection in the Renaissance and throughout the Enlightenment, grave robbery—or body snatching—flourished and supplied an increasing amount of fresh corpses for anatomical teaching in the same period.[73] In many countries, executed criminals were handed over for dissection with the disruption of the corpse probably considered a fair part of the punishment,[74] and at the end of the eighteenth century, the guillotine delivered corpses in abundance for François-Xavier Bichat's research in France.[75] Though representing different political games enrolling different bodies in knowledge production, the point is that not every family has been capable of taking care of their deceased.[76] It differs which bodies serve as means for the needs of others; but divisions in relative worth and status appear to be a relatively consistent element of social history.

Entitlements of the Living: The Rise of Agential Giving

The Anatomy Act of 1832 was supposed to put an end to grave robbery in the UK, but as pointed out by Naomi Pfeffer, it also facilitated the opportunity of *donating* one's body to the medical profession. Again the subject (as a decision maker) became separated out from the body (as materiality). In the nineteenth century, a movement of intellectuals in France called the Society of Mutual Autopsy actively began propagating the idea that bodies—through dissection—should be put to public use after death. This time it was an elite actively propagating body donation and even offering their own.[77] The notion that dissection was an altruistic act decided upon by an enlightened individual, rather than a capital punishment sanctioned by the sovereign, emerged in tandem with the changing political landscape described above where the sovereign no longer ruled over bodies but conducted the conduct of autonomous individuals. Some bodies, of course, remained more obliged than others to further the public good. Well into the twentieth century, some countries had statutes according to which the bodies of people who could not afford their own funeral should be handed over to the medical profession.[78] The emergence of a state purview created a new realm within which dead bodies could be enrolled into knowledge production without provoking the type of riots against body snatching that had led to statutes declaring to stop trade in corpses. The corpse of the average citizen was more and more effectively withdrawn from certain types of trade (such as body snatching), but corpses were not withdrawn from exchange. They were made into objects of donation. Contestation of body trade thereby facilitated a more seamless harvesting model and secured a steadier flow of corpses.

If the dead body (and its ubjects) was generally excluded from legitimate trade before the living body of the slave came to experience this privilege, it is yet another story what happened to *ubjects* taken from living bodies. Ubjects taken from patients admitted to hospital were generally seen as *abandoned waste* and of no interest to the patient. Excised limbs, tissue, or examples of deformities were readily collected, used in teaching, and exhibited. Anatomical collections such as Galli's in Bologna, Hunter's in London, Saxtorph's in Copenhagen, and Mütter's in Philadelphia exhibit the confidence with which anatomists, surgeons, and doctors have collected specimens of special interest to their profession. Well into the twentieth century, excised tissue has been appropriated by medical professionals for differing purposes, as also the Alder Hey scandal of retained children's organs in the UK exemplifies.[79]

Some living donors were very well aware of the ubjects they delivered. When Kant took teeth as an example of what one should not sell (see Chap. 1), it was not an arbitrary example. Live tooth transplantation was a thriving surgical specialty in the eighteenth century, and only when it came to be associated with transmission of otherwise sexually transmitted diseases did boys and girls from the lower classes stop relinquishing their teeth to the sugar-inflicted higher classes.[80] Mark Blackwell describes how the practice was contested not only because it deprived the poor of health for money but because it involved multiple transgressions and served nothing but disgraceful goals of vanity (and thereby a lack of self-restraint). In the early nineteenth century, wealthy Londoners would import their teeth from what became deemed a safer source, namely, teeth taken from dead soldiers, all rinsed and conveniently packaged. They became known as Waterloo teeth named after a place in which they could be found in unlucky abundance during those years.[81] Some corpses were apparently available for consumption by wealthy recipients irrespective of the *nullius in bonis* principle, at least when the ubjects of living bodies were not seen as an appealing option and as long as the corpse was from far away.

The uncertainties surrounding what can be collected from whom relate to the more basic ambiguity concerning what is part of whom and what constitutes a body. Stillborns with rare deformities were readily collected in all the collections mentioned above and apparently not regarded as *nullius in bonis*. They were not "corpse" and not "part of" the living mother. They seem to have been viewed as a sort of (valuable) waste at the disposition of the doctor in accordance with the "abandoned waste" rule. This principle also became contested, however, and in particular so when doctors used such collections commercially. From the 1740s, court cases document instances where parents question the doctor's right to exhibit their stillborns or use excised surgical material.[82] In cases like these, entitlements are negotiated, thereby exhibiting a changing moral landscape in which profit from ubjects is becoming increasingly conspicuous. The moral paradigm manifests itself through such contestations, and gradually we have seen a shift toward exchange forms where detachment of ubject from subject becomes legitimate mainly when sanctioned by the type of possessive, autonomous actor described above, which made little sense pre-enlightenment.

Contemporary Contestations

If we move to more recent examples of body/commerce interfaces often dis-cussed in the literature, we find examples of individuals feverously *trying* to make a profit out of ubjects from their bodies and not just families trying to avoid this type of profit-making. Almost as a sequel to the tooth-selling of the eighteenth and nineteenth centuries, blood plasma production involves inviting people to donate for money in some countries, the USA included. For a number of reasons, there is an international move toward avoiding monetary compensa-tion in these industries, as mentioned in Chap. 1, even if the voice of market proponents cannot be said to cease. New technologies in fertility treatment have intensified the desire for gametes (because, as Harvey noted, technologies are part of determining what constitutes resources), and in the USA this has made both men and women try to cash in on the gametes that their bodies produce. In most countries it is seen as illegal to *sell* these cells; so donors are *compensated*. This is an interesting distinction I will explore in more depth in Chap. 4. In some European countries, compensation of women is more strictly regulated than that of men, and in effect many sperm donors do see themselves as being paid, though not always listing the payment as the primary or only reason for their donation.[83]

Interestingly, it was also a case pertaining to sperm that in 2009 first made a British court provide a man with property rights to his body. The case did not revolve around commercial transactions, but was about negligence on behalf of a hospital storing frozen sperm from cancer patients and letting it thaw so that it lost its usability. The court ruling specifies that the sperm is not *part of* the body, but that "property" is such a broad concept that one can in cases like this single out some property rights (right to use the sperm) and see them violated through negligence.[84] Persons opting for commercial rights to sell their body parts have not been granted property rights in court, but sperm is probably the ubject closest to commodity status and it will be interesting to see if sperm donors will use the court ruling at some point to claim ownership of the ubjects they produce. In 2005, Portugal passed a law in which tissue was designated property of the donor, but again it seems to use the word property to refer to entitlements of disposal, not commercial gain.[85] Similarly, an American court used the notion of property in relation to "body parts" already in 1891, again without aligning property rights with the commercial aspects associated with market thinking and without this ruling having had any subsequent influence.[86]

Unlike the destitute tooth-sellers and the more recent blood and gamete donors, another type of individual opting for profit is the research subject acting on knowledge that they gain more or less by coincidence. The most famous is probably John Moore, who in 1983 learned that his doctor had produced a cell line out of material taken from his spleen during cancer surgery in 1976.[87] Moore had gone for several follow-up visits not knowing that they facilitated the doctor's research. He sued for conversion to gain a percentage of the expected million dollar profit from the cell line. Moore lost the case in a seminal ruling which stipulated the patient's entitlement to informed consent, though not to profit. According to the majority vote, profit belonged in the hands of the researchers who applied their skills to make cells into a cell line, not the person delivering the cells. The moral paradigm thereby came to manifest itself as a bifurcation of bodily substance and research skill (see Chap. 4). Another man, more successful at capitalizing on his cells, was Ted Slavin, who (unlike Moore) was informed by his doctor that his blood cells were quite rare. As a hemophiliac growing up in the 1950s, he had been exposed numerous times to hepatitis B virus, and his blood therefore contained unusually high concentrations of hepatitis B antibodies. He began traveling around research laboratories selling samples of blood to researchers working on hepatitis B, but he also singled out a doctor he trusted to do particularly valuable research and volunteered as research subject for him. Even for Ted Slavin, the body was not just a source of income. Slavin also founded a company (Essential Biologicals) for similarly "endowed" individuals.[88] Even in wealthy countries, there are isolated examples also of people trying to sell their organs.[89]

When we contemplate these cases of ubjects entering monetary exchange schemes, we should take care not to extrapolate from singular instances to the motivation of all potential donors. There is not one set of interests characterizing all individuals opting for income from their body, and more importantly, I think individuals will have differing senses of control in the various situations they are bringing themselves into. The following chapters will go into more depth with the sense of control arguing that it is of key importance for how people relate to ubjects.

Power, Resistance, and Morality

In practically all the contestations listed above, *money* as a form of power potentially loses some of the legitimacy, which it holds in other domains, when it is used
to deprive others of health or control over their body. The case I wish to make is that
in the various forms of contestation, we see a form of resistance to monetary power.
While the state buttresses the power of money in some domains, a new morality
emerges which sets limits to this power, and it is not unlike what Bloch and Parry
referred to as balancing. These limits help lifting bodies out of the realm of commodities. Thomas Haskell has suggested that humanitarianism is a good owed to the
rise of capitalism and that we might stand to lose it. Equally poignant, we might
have to consider what the bioeconomy will do to the moral protection of bodies that
materialized with the rise of market thinking.

I will return to this question in the conclusion. Here, I will end instead with a few
reflections on the more immediate question about how we might understand the fact
that various ubjects—from dead and from living bodies—continue to become
enrolled into exchange systems despite continual contestation from many quarters.
It is tempting to suggest there are no definitive moral rules that cannot be circum-
vented when it is apt for people who have the means for pursuing their ends. And
certainly there is some truth to this. One should be careful however not to reduce the
agency going into contestations of this type into a struggle between agents driven by
ruthless power (trying to acquire ubjects) countered by morally informed resistance
(trying to defend ubjects). It is far too simple, and it would in effect imply seeing
most of today's market proponents as ruthless agents of power. Clearly, they would
not subscribe to such a crude reduction of their moral engagement. And indeed
I think it also gives little credit to the fact that people using the "I should be allowed
to do what I can afford to do" argument in relation to ubject exchange actually pur-
port moral ideals of justice associated with market thinking and often speak on
behalf of potential recipients they do not even know. Furthermore, as one would
expect of an analysis using a Foucauldian notion of resistance, I think we should see
the contestations as elements in the shaping of a wider power game. The moral
legitimacy that the distinction between bodies and commodities creates is part of
expanding the legitimacy of market thinking. It is through contestations that entitlements to ubjects have gradually acquired the shape they have today and through
them that the moral paradigm of market thinking has gained threshold. With it, other
aspects of market thinking becomes ingrained. When, for example, a number of
court cases from mid-eighteenth century onward establish a problem with selling
anatomical collections, unless it can be construed as selling something else (e.g., the
jars and the work going into the preparation of the specimen),[90] it contributes to the
development of new moral landscapes both curtailing *and* expanding what money
can be legitimately used for. Similarly, the rise of a patent regime which implies that
ubjects are not owned, but knowledge about them is, facilitates both "protection" of
the ubject and expansion of another form of power, the patent, which is at a safe
distance from the average donor as well as the individual researcher. In all of these
cases, we see a society guarding its own borders, determining what is and is not part

of the community it protects.[91] This attendance to borders reflects the distinctions—described above in relation to Fig. 2.1—between nature and culture, things and citizens, commodities and persons.

The overall point I have made is that during these multiple historical transformations, commodities acquired traits construed almost as the counterimage of the emerging ideas about the human person: they came to appear nonhuman, passive, in no need of respect, and possibly enriched but never enriching per se. They became the natural means to the human person as an end. This moral paradigm can today be seen as posing particular problems when ubjects are exchanged—but not all ubjects, not at all times, not to every form of exchange. And the paradigm only takes the form that it does due to hybrids stimulating the purification work. Because of their hybridity, ubjects stimulate the work through which persons and things come to be viewed as essentially incompatible.

Exchange as a Set of Evolving Practices

There is no way to reach agreement about what a market is, and yet this concept holds incredible salience in debates about practically everything else. This chapter has outlined the central components of what I call *market thinking* and a corresponding set of exchange practices. "Markets" and "market forces" are not empirical facts abiding to natural laws in the sense often referred to in the debate about "markets in human body parts." Market thinking is a historically specific way of making sense of historically specific exchange practices. And these practices change. I have shown how scholars from anthropology and history have developed alternative conceptualizations, which can be helpful in understanding the specificity of the moral ideals now associated with "markets." The intertwined social and intellectual history portrays the rise of a biopolitical configuration that cannot be reduced to either material practice or ideas as if one was the cause of the other.

As presented above, it is very much an Anglophone history. The analytical point, however, is more general and not so much concerned with identifying the historical origin of capitalism or the like.[92] My point is that the current contestation of ubjects in biomedical practices indicates a continuous interplay of modes of production, of forms of knowledge, of notions of virtue and personhood, and of links between personhood and bodies. I have wanted to show that the moral paradigm currently making particular ubjects seem controversial has not always been in place; it was coproduced with the current mode of production in which ubjects now perform many types of work, which for some appear uncanny. The moral exemption of "bodies" from commercial trade was never given, and as a contingent historical product, it can disappear again. The proliferation of controversies revolving around ubjects is indicative of new interfaces between the mode of production, the medical field, and dominant moral norms, and it is through these controversies we currently negotiate publicly what is part of a body. But before we get too upset with finding out what is "part," we should reflect a little more on what we consider to be "whole." What is a body, really?

Endnotes

1. Lene Koch's work on the changing meanings of eugenics has been a source of great inspiration for me in this respect (Koch 1996, 2004).
2. Marx (1978:155).
3. Callon (1998), Callon and Muniesa (2005), and Carrier (1997).
4. See also Dilley (1992).
5. A stark market proponent, Mark Cherry, finds justification for this view in Locke and Kant and finds in it basic moral value. He also suggests that ubjects can be seen as things (Cherry 2005:chapter 2).
6. Harvey (2001).
7. Kant (1997:157).
8. A number of initiatives have been taken to protect the human body from trade, repeatedly making material derived from human bodies subject to special rules. International conventions, such as the Council of Europe's Convention on Human Rights and Biomedicine and UNESCO's Universal Declaration on the Human Genome and Human Rights, explicitly prohibit financial gain from the human body and its parts in their natural states. Numerous forms of national legislation have been put in place to ensure that bodily products cannot be commercially purchased from donors. Even in places where the exchange of body parts is described as a literal trade, this trade is either considered illegal theft or a black market business (Kovac 1998), or it represents state authorities using extensive powers of an altogether noncapitalist kind as in China (Becker 1999), or it is officially illegal but subject to a business specializing in obscuration as with kidney transfers in India which are made to appear as donations between friends (Muraleedharan et al. 2006). It can also be obscured as "processing fees" as in the American circulation of tissue from cadavers to the cosmetic industry (Timmermans 2006:243), as I will discuss in more depth in Chap. 4. Iran is, of course, a notable example where organs are procured with monetary means and the state's blessing, but I think we should be careful not to assume that we know the mechanisms of exchange even if we decide to call the Iranian case a "market."
9. Caliskan and Callon (2009) and Zelizer (1979).
10. Mason and Laurie (2001:715). See Chap. 1.
11. Wilde (2003).
12. Geismar (2001).
13. Zelizer (2011:ix). See also Zelizer (1979, 1998).
14. Friedl and Robertson (1990) and Hart (1990). See also Foucault's analysis of the rise of economics (Foucault 1999). Jack Goody (2006) finds such claims Eurocentric and indicative of lacking familiarity with Asian history. I am concerned here with showing how ideas about persons, bodies, moral entitlements, power structures, and the mode of production are coproduced rather than with the origination of particular concepts per se, and I will focus mostly on ubject exchange in Europe and North America.
15. For a Marxist-feminist critique of the way in which "economy" has acquired a paradoxical role as both dictating political decisions out of necessity and constituting the prime arena for judgment as political choice, see Gibson-Graham (2006).
16. Economics is a diverse field with multiple and competing ways of understanding how and why people organize their exchanges in various ways. By granting Elinor Ostrom and Oliver Williamson the Nobel Prize in Economics 2009, the Nobel Committee illustrated the heterogeneity of economics and its blurred boundaries to other social science disciplines. Especially Ostrom's work on user-driven domains questions some of the assumptions just characterized as typical for market thinking.
17. Hart (1990). Many economic theories have proven themselves very useful for understanding and sometimes even predicting how populations respond to various stimuli. The extent to which economics revolves around self-fulfilling prophecies continues to be disputed.
18. Scott (1976). See also Sivaramakrishnan (2005).
19. Polanyi (1957). Fredrik Barth (1967) is an example of a prominent formalist.

20. A literature with many parallels developed in economic sociology, in particular following
    Zelizer's pathbreaking work on the emergence of the life insurance industry (Zelizer 1979). It
    lies beyond the limits of this chapter to capture and synthesize the scholarship dealing with
    these issues in economics, sociology, and anthropology. I focus on the anthropological debate
    because it satisfyingly brings out the points needed for my argument about the coproduction of
    social categories and exchange.
21. Bohannan and Bohannan (1968).
22. Some market proponents in the organ debate explicitly state that a way to introduce markets in
    a publically acceptable manner could be to introduce only incentives seen as belonging to the
    "same sphere," that is, health insurance (Beierholm et al. 2009).
23. Latour and Lépinay (2009).
24. Barth (1967).
25. Appadurai (1986) and Kopytoff (1986).
26. Eiss and Pedersen (2002).
27. Bloch and Parry (1989).
28. This point has been made by many scholars; see, for example, the commentary on Koselleck
    (Ifversen 2003, 2007). The notion of looping effects is borrowed from Ian Hacking (1995).
29. Michel Callon refers to his own theoretical approach as the performativity program (Caliskan
    and Callon 2009; Callon 1998). For an introduction to similar developments in geography, see
    Lee et al. (2008).
30. Caliskan and Callon (2009). See also Callon and Muniesa (2005).
31. Lianna Farber (2006), for example, warns against looking for medieval writings about econ-
    omy and measuring them against today's standards. Rather, we should decide on particular
    parameters associated with exchange and see how they were debated at the time. She suggests
    that medieval writers associated exchange with (1) assessments (value), (2) measures for
    reaching agreement (consent), and (3) ideas about how to live together (community). These
    were topics of great importance to people, but they formed part of a very different moral land-
    scape than that carved out by market thinking.
32. See, for example, how Scheper-Hughes describes the genealogy of commoditization as ranging
    "from the animated sale, collection and veneration of medieval relics of the bodies of Catholic
    saints to the grave-robbings of the 16th and 17th centuries by barbers and surgeons in search of
    corpses for dissection and for teaching gross anatomy" (Scheper-Hughes 2001:3).
33. Lawrence (1998:116).
34. Esmark (2002).
35. Mauss (2000:79).
36. Marx (1972). The Scandinavian word "vare" is derived from German.
37. Ibid.:320–321.
38. Marx (1972:303).
39. In this sense, Marx focused on economic rather than moral alienation as loss of integrity (as it
    is often the case in debates about body parts). It was, of course, still a morally infused critique
    concerning what *ought* to belong to the worker.
40. Weber (1992).
41. Haskell (1985). Haskell has been criticized for not acknowledging class interests in his expla-
    nation of the historical developments leading to the abolition of slavery (Ashworth 1992;
    Bender 1992). However, I am not interested in the role of class and the debate about false
    consciousness here. Rather, I aim to point out that a new mode of production is intertwined
    with the emergence of new moral concerns.
42. Bill Brown has argued that slavery did not involve seeing slaves only as things; in fact part of
    the gruesomeness was the conscious confusion between thing and person through which
    persons (not things) became exploited for the pleasure of others (Brown 2006). The rise of a
    sharp distinction between persons and things is a product of purification work which intensified
    during and after the abolition with a new mode of production and consumption.
43. Gold (1996:6).
44. Macfarlane (1998).

45. See also Weiner (1992).
46. For differences between the UK, France, and Germany in the changes of translating this type of power into political influence, see Elias (1994a, b).
47. See Witte and Have (1997).
48. Foucault (1999).
49. Pocock (1985).
50. Pocock (1985:114).
51. See also Waldby's interesting discussion of the imagery of regenerative medicine as a means for achieving ultimate autonomy: the ability to generate yourself without father and mother (Waldby 2002:316).
52. Kant (1997:126). This change in the conception of autonomy has been explicitly promoted in some quarters of philosophy. A strong market proponent like James Stacey Taylor, for example, has also published in favor of reinterpreting autonomy as "endorsed preference" devoid of its earlier connotations to self-restraint (Taylor 2005).
53. This image has been contested long before ANT emerged. Heidegger is famous for pointing out how development of technologies for exploiting the material world has a looping effect on human beings too; technologies emerge through material practices in which persons are worked upon and not just working upon (Heidegger 1999). Also Friedrich Engels pointed out that technologies tend to take revenge for any attempt of controlling "nature" by turning the control around and subjecting workers to authority (Engels 1976).
54. Elias (1994a).
55. See also Rose (1992, 1999).
56. Foucault (2002).
57. Foucault (1991).
58. Think also of Norbert Elias' description of the interdependence of the rise of taxation, state power, and self-restraint (Elias 1994a).
59. At the same juncture, the UK opted for a liberal approach under strong public oversight in relation to embryonic stem cell research, while several southern European countries tried to curtail the research field as a whole.
60. Obama's initial attempt of lifting Bush's ban in 2010 was first said to be unlawful with reference to a 1996 congressional amendment stating that embryos may not be destroyed in the course of federally funded research (Blackburn-Starza 2010).
61. Thacker (2005); see also Cooper (2006, 2007), Fortun (2008), and Rajan (2003, 2006).
62. Post-industrialization is also associated with changing power relations through which persons are expected to regard themselves as malleable and ready to adopt any agenda which is profitable for their employer (Adkins 2005). Deleuze (1990) suggests that the resulting transfigurations of power are more pervasive than what Foucault described.
63. The expression of "life itself" has been used by a number of prominent scholars including Nikolas Rose (2007). I find it striking how it remains strangely vague and somehow at odds with a focus among the same scholars on socio-material coproduction (Is "life itself" somehow distinct from the social world?), and yet it provokes a strong feeling of urgency. The popularity of the term "life itself" in relation to processes of capitalization is in my view indicative of the way the moral paradigm associated with market thinking described above tacitly influences academic priorities through evocative prose.
64. Thacker (2005:204).
65. In Thompson's (2005) work on the biomedical mode of production, a similar tendency to highlight biological work as special for the current mode of production can be traced, though I think it could be much clearer what distinguishes the capitalization on the work of human cells from that of, for example, animals (Shukin 2009; Wolfe 2010). See also Pálsson and Harðardóttir (2002).
66. Cooper (2008:147); see also Parry (2004).
67. Hoeyer (2007).
68. Boyer and Cohen published their techniques for producing recombinant DNA in 1973 and based on patents for this technique founded Genentech in 1980 (Thacker 2005:173–174).

69. Nevertheless, it is quite common to see grand claims about the political history of the body which actually ought to poke more questions of why this type of narrative has such appeal. Some scholars, for example, have placed great emphasis on the Habeas Corpus Act of 1679 in Great Britain and claim that it illustrated the need for a body in which to locate the citizen as subject of power (Cohen 2008). The type of statute has been worked into legal systems in many countries. Giorgio Agamben also claims habeas corpus as a central biopolitical moment linking law directly to life (Agamben 1995:123). Habeas corpus specified that detained persons were entitled to be presented—as bodies—in court.

70. It might appear provocative to draw an analogy to Rayna Rapp's (1988) notion of moral pioneers, but I think there are clear similarities between people trying to make sense of new technologies and people trying to do new things to bodies. When, for example, William Hunter in the eighteenth century encouraged body snatching, he pioneered a new morality in approaches to the body as an object of scientific investigation; see Moore (2005).

71. U.S. Congress (1987). Jackson claims that common law always had some influence and that the strong position of the church was limited to a short period in the ninth to thirteenth centuries (Jackson 1937).

72. Skegg (1975). *Nullius in bonis* can also be taken to mean "for the benefit of no person," and though Skegg is not aware of its origin, classical scholar Kirsten Jungersen (p.c.) has kindly pointed out to me that it derives from Caius (second century A.D.).

73. See the vivid descriptions in Moore's fascinating biography on the anatomist and natural scientist John Hunter (Moore 2005). Students of art were also known to steal corpses in medieval and Renaissance Europe.

74. Anatomical dissections on the bodies of executed criminals were a significant source of income (and fame) for Europe's oldest university in Bologna already in the 1300s (Ceglia 2006; Ferrari 1987). Interestingly, the Italian city states were in some respects forerunners to many of the trade aspects of the market model, though not the industrial mode of production.

75. Pfeffer (forthcoming).

76. On the emotional aspects of caretaking of dead relatives as a material practice, see Langford (2009).

77. Green (2008) and Hecht (2003:44).

78. In Sweden, for example, it was in principle the case into the 1960s (Åkesson 1996).

79. Clarke (1995), Landecker (2007), and Morgan (2002). On Alder Hey's representation in the media, see Seale et al. (2006).

80. Blackwell (2004).

81. Ibid.

82. Skegg (1975, 1992).

83. Sebastion Mohr (p.c.), Steiner (2006). Egg donors in various places have also been shown to be motivated by more than the money (Motluk 2010; Orobitg and Salazar 2005). Almeling (2007) also shows that in the USA the price does not just reflect some abstract form of supply/demand mechanisms but assessments of the suitability of buyers so that same-sex couples pay higher prices for gametes. A somewhat methodologically problematic survey of 485 young adults born by sperm donors came up with the quite interesting result that 42% of them thought it "wrong for people to provide their sperm or egg for a fee" (Marquardt et al. 2010:84).

84. Yearworth and others v. North Bristol NHS Trust (2009).

85. The impact of the law is yet to be tested in court (de Faria 2009).

86. U.S. Congress (1987).

87. Moore v. Regents of the University of California. Many people have analyzed this case; see in particular Gold (1996), Landecker (1999), and Rabinow (1992).

88. The story of Ted Slavin is forcefully recounted in Rebecca Skloot's (2010) beautifully written bestseller about Henrietta Lacks.

89. Think of the occasional eBay auction. I will discuss surveys indicating vendor interest in this field in Chap. 4 (Jasper et al. 1999). See also Friedlaender (2002) on organ sale in Israel.

90. Ibid.; it is in these cases that the no property rule is established but, according to Skegg, with a misconstrued understanding of its legal grounds.

91. I would like to thank philosopher Hans Ruin for making me think more clearly about this point (personal communication).
92. Featherstone (2009).

References

Adkins L (2005) The new economy, property and personhood. Theory Cult Soc 22(1):111–130

Agamben G (1995) Homo Sacer: sovereign power and bare life. Stanford University Press, Stanford

Åkesson L (1996) The message of dead bodies. In: Lundin S, Åkesson L (eds) Bodytime: on the interaction of body, identity, and society. Lund University Press, Lund, pp 157–180

Almeling R (2007) Selling genes, selling gender: egg agencies, sperm banks, and the medical market in genetic material. Am Sociol Rev 72(3):319–340

Appadurai A (1986) Introduction: commodities and the politics of value. In: Appadurai A (ed) The social life of things: commodities in cultural perspective. Cambridge University Press, Cambridge, pp 3–63

Ashworth J (1992) The relationship between capitalism and humanitarianism. In: Bender T (ed) The Antislavery debate: capitalism and abolitionism as a problem in historical interpretation. California University Press, Los Angeles, pp 180–199

Barth F (1967) Economic spheres in Darfur. In: Firth R (ed) Themes in economic anthropology. Tavistock, London

Becker C (1999) Money talks, money kills—the economics of transplantation in Japan and China. Bioethics 13(3/4):236–243

Beierholm ML, Platz TT, Østerdal LP (2009) Organdonation og -allokering i et mikroøkonomisk perspektiv. Nationaløkonomisk Tidsskr 147(3):265–296

Bender T (1992) Introduction. In: Bender T (ed) The Antislavery debate: capitalism and abolitionism as a problem in historical interpretation. University of California Press, Los Angeles, pp 1–13

Blackburn-Starza A (2010) US judge rules against Obama's stem cell policy. Bionews, p 573

Blackwell M (2004) "Extraneous bodies": the contagion of live-tooth transplantation in late-eighteenth-century England. Eighteenth-Century Life 28(1):21–68

Bloch M, Parry J (1989) Introduction: money and the morality of exchange. In: Parry J, Bloch M (eds) Money and the morality of exchange. Cambridge University Press, Cambridge, pp 1–32

Bohannan P, Bohannan L (1968) Tiv economy. Northwestern University Press, Evanston

Brown B (2006) Reification, reanimation, and the American uncanny. Crit Inq 32(2):175–207

Caliskan K, Callon M (2009) Economization, part 1: shifting attention from the economy towards processes of economization. Econ Soc 38(3):369–398

Callon M (1998) The embeddedness of economic markets in economics. The laws of the markets. Blackwell Publishers, Oxford, pp 1–57

Callon M, Muniesa F (2005) Economic markets as calculative collective devices. Organ Stud 26(8):1229–1250

Carrier JG (1997) Introduction. In: Carrier JG (ed) Meanings of the market: the free market in western culture. Berg, Oxford, pp 1–67

Ceglia FD (2006) Rotten corpses, a disembowelled woman, a flayed man. Images of the body from the end of the 17th to the beginning of the 19th century. Florentine wax models in the first-hand accounts of visitors. Perspect Sci 14(4):417–456

Cherry MJ (2005) Kidney for sale by owner: human organs, transplantation, and the market. Georgetown University Press, Washington, DC

Clarke AE (1995) Research materials and reproductive science in the United States, 1910–1940. In: Star SL (ed) Ecologies of knowledge. State University of New York, Albany, pp 183–225

Cohen E (2008) A body worth having? Or, a system of natural governance. Theory Cult Soc 25(3):103–129

Cooper M (2006) Resuscitations: stem cells and the crisis of old age. Body Soc 12(1):1–23
Cooper M (2007) Life, autopoiesis, debt: inventing the bioeconomy. Distinktion 14:25–43
Cooper M (2008) Life as surplus: biotechnology and capitalism in the neoliberal era. University of
    Washington Press, Seattle
de Faria P (2009) Ownership rights in research biobanks: do we need a new kind of 'biological
    property'? In: Solbakk JH, Holm S, Hofmann B (eds) The ethics og research biobanking.
    Springer, New York
Deleuze G (1990) Kontrol og tilblivelse. In: Deleuze G (ed) Forhandlinger 1972–1990. Det lille
    forlag, Frederiksberg, pp 203–220
Dilley R (1992) Contesting markets: a general introduction to market ideology, imagery and dis-
    course. In: Dilley R (ed) Contesting markets: analyses of ideology, discourse and practice.
    Edinburgh University Press, Edinburgh, pp 1–34
Eiss PK, Pedersen D (2002) Introduction: values of value. Cult Anthropol 17(3):283–290
Elias N (1994a) State formation and civilization: the civilizing process. Blackwell, Oxford, pp
    257–544
Elias N (1994b) The history of manners: the civilizing process. Blackwell, Oxford, pp vii–255
Engels F (1976) Om Autoritet. Marx/Engels udvalgte skrifter I. Forlaget Tiden, København, pp
    632–636
Esmark K (2002) De Hellige Døde og Den Sociale Orden: Relikviekult, Ritualisering og Symbolsk
    Magt. University Center of Roskilde, Roskilde
Farber L (2006) An anatomy of trade in medieval writing: value, consent, and community. Cornell
    University Press, Ithaca
Featherstone M (2009) Occidentalism: Jack Goody and comparative history: introduction. Theory
    Cult Soc 26(7–8):1–15
Ferrari G (1987) Public anatomy lessons and the carnival: the anatomy theatre of Bologna. Past
    Present 117:50–106
Fortun M (2008) Promising genomics. University of California Press, Berkeley
Foucault M (1991) Governmentality. In: Burchell G, Gordon C, Miller P (eds) The foucault effect:
    studies in governmentality. The University of Chicago Press, Chicago, pp 87–104
Foucault M (1999) Ordene og Tingene. En Arkæologisk Undersøgelse af Videnskaberne om
    Mennesket [The order of things]. Spektrum, Viborg
Foucault M (2002) Overvågning og Straf [Dicsipline and punish]. Det Lille Forlag, Copenhagen
Friedlaender MM (2002) The right to sell or buy a kidney: are we failing our patients? Lancet
    359(9310):971–973
Friedland R, Robertson AF (1990) Beyond the martketplace. In: Friedland R, Robertson AF (eds)
    Beyond the marketplace: rethinking economy and society. Aldine de Gruyter, New York, pp 3–13
Geismar H (2001) What's in a price? An ethnography of tribal art at auction. J Mater Cult
    6(1):25–47
Gibson-Graham JK (2006) The end of capitalism (as we knew it): a feminist critique of political
    economy. University of Minnesota Press, Minneapolis
Gold ER (1996) Body parts: property rights and the ownership of human biological materials.
    Georgetwon University Press, Washington, DC
Goody J (2006) The theft of history. Cambridge University Press, New York
Green JW (2008) Beyond the good death: the anthropology of modern dying. University of
    Pennsylvania Press, Philadelphia
Hacking I (1995) The looping effects of human kinds. In: Sperber D, Premack D, Remack AJ (eds)
    Causal cognition: a multidisciplinary debate. Clarendon, Oxford, pp 351–394
Hart K (1990) The idea of economy: six modern dissenters. In: Friedland R, Robertson AF (eds)
    Beyond the marketplace: rethinking economy and society. Aldine de Gruyter, New York, pp
    137–160
Harvey D (2001) Population, resources, and the ideology of science. In: Spaces of capital: towards
    a critical geography. Edinburgh University Press, Edinburgh
Haskell TL (1985) Capitalism and the origins of the humanitarian sensibility, Part 2. Am Hist Rev
    90(3):547–566

Hecht JM (2003) The end of the soul: scientific modernity, atheism, and anthropology in France. Columbia University Press, New York

Heidegger M (1999) Spørgsmålet om Teknikken [Die Frage nach der Technick]. In: Goll J, Zahavi D (eds) Spørgsmålet om Teknikken og Andre Skrifter. Samlerens Bogklub, Copenhagen, pp 36–65

Hoeyer K (2007) Person, patent and property: a critique of the commodification hypothesis. BioSocieties 2(3):327–348

Ifversen J (2003) Om den tyske begrebshistorie. Politologiske Studier 6(1):18–34

Ifversen J (2007) Begrebshistorien efter Reinhart Koselleck. Slagmark 48:81–103

Jackson PE (1937) The law of cadavers and of burial and burial places. Prentice-Hall, Inc., New York

Jasper JD, Nickerson CAE, Hershey JC, Asch DA (1999) The public's attitudes towards incentives for organ donation. Transplant Proc 31:2181–2184

Kant I (1997) Lectures on ethics: Immanuel Kant. Cambridge University Press, Cambridge

Koch L (1996) Racehygiejne i Danmark 1920–56 [Eugenics in Danmark 1920–56]. Gyldendal, Copenhagen

Koch L (2004) The meaning of eugenics: reflections on the government of genetic knowledge in the past and the present. Sci Context 17(3):315–331

Kopytoff I (1986) The cultural biography of things: commoditization as process. In: Appadurai A (ed) The social life of things: commodities in cultural perspective. Cambridge University Press, Cambridge, pp 64–91

Kovac C (1998) Tissue trade in Hungary is investigated. BMJ 316:645

Landecker H (1999) Between beneficence and chattel: the human biological in law and science. Sci Context 12(1):203–225

Landecker H (2007) Culturing life: how cells became technologies. Harvard University Press, Cambridge

Langford JM (2009) Gifts intercepted: biopolitics and spirit debt. Cult Anthropol 24(4):681–711

Latour B, Lépinay VA (2009) The science of passionate interests: an introduction to Gabriel Tarde's economic anthropology. Prickly paradigm press, Chicago

Lawrence S (1998) Beyond the grave—the use and meaning of human body parts: a historical introduction. In: Weir RF (ed) Stored tissue samples: ethical, legal, and public policy implications. University of Iowa Press, Iowa City, pp 111–142

Lee R, Leyshon A, Smith A (2008) Rethinking economies/economic geographies. Geoforum 39:1111–1115

Macfarlane A (1998) The mystery of property: inheritance and industrialization in England and Japan. In: Hann CM (ed) Property relations. Cambridge University Press, Cambridge, pp 104–123

Marquardt E, Glenn ND, Clark K (2010) My daddy's name is doner: a new study of young adults conceived through sperm donation. Institute for American Values, New York

Marx K (1972) Capital. In: Tucker R (ed) The Marx-Engels reader. W.W. Norton & Co., New York, pp 294–437

Marx K (1978) The German ideology. In: Tucker R (ed) The Marx-Engels reader. W.W. Norton & Company, London, pp 146–200

Mason JK, Laurie GT (2001) Consent or property? Dealing with the body and its parts in the shadow of Bristol and Alder Hey. Mod Law Rev 64(5):710–729

Mauss M (2000) The gift: the form and reason for exchange in archaic societies. Routledge, London

Moore W (2005) The knife man. Bantam Press, London

Morgan L (2002) "Properly disposed of": a history of embryo disposal and the changing claims on fetal remains. Med Anthropol 21(3):247–274

Motluk A (2010) The human egg trade: how Canada's fertility laws are failing donors, doctors, and parents. The Walrus 7(3):30

Muraleedharan VR, Jan S, Prasad SR (2006) The trade in human organs in Tamil Nadu: the anatomy of regulatory failure. Health Econ Policy Law 1:41–57

Orobitg G, Salazar C (2005) The gift of motherhood: egg donation in a Barcelona infertility clinic. Ethnos 70:31–52

Pálsson G, Harðardóttir K (2002) For whom the cell tolls. Curr Anthropol 43(2):271–301

Parry B (2004) Bodily transactions: regulating a new space of flows in "bio-information". In: Verdery K, Humphrey C (eds) Property in question: value transformation in the global economy. Berg, Oxford, pp 29–68

Pfeffer N (forthcoming) Insider trading. Yale University Press, London

Pocock JGA (1985) The mobility of property and the rise of eighteenth-century sociology. In: Pocock JGA (ed) Virtue, commerce, and history: essays on political thought and history, chiefly in the eighteenth century. Cambridge University Press, New York, pp 103–123

Polanyi K (1957) The great transformation. Beacon, Boston

Rabinow P (1992) Severing the ties: fragmetation and dignity in late modernity. Knowl Soc Anthropol Sci Technol 9:169–187

Rajan KS (2003) Genomic capital: public cultures and market logics of corporate biotechnology. Sci Cult 12(1):87–121

Rajan KS (2006) Introduction: capitalisms and biotechnologies. Biocapital: the constitution of postgenomic life. Duke University Press, London, pp 1–36

Rapp R (1988) Moral pioneers: women, men and fetuses on a frontier of reproductive technology. In: Baruch E, D'Adamo A, Seager J (eds) Embryos, ethics, and women's rights: exploring the new reproductive technologies. Haworth Press, New York City

Rose N (1992) Towards a critical sociology of freedom. University of London, London

Rose N (1999) Powers of freedom: reframing political thought. Cambridge University Press, Cambridge

Rose N (2007) The politics of life itself: biomedicine, power, and subjectivity in the twenty-first century. Princeton University Press, Princeton

Scheper-Hughes N (2001) Bodies for sale—whole or in parts. Body Soc 7(2–3):1–8

Scott J (1976) The moral economy of the peasant: rebellion and subsistence in Southeast Asia. Yale University Press, London

Seale C, Cavers D, Dixon-Woods M (2006) Commodification of body parts: by medicine or by media? Body Soc 12(1):25–42

Shukin N (2009) Animal capital: rendering life in biopolitical times. University of Minnesota Press, Minneapolis

Sivaramakrishnan K (2005) Introduction to "moral economies, state spaces, and categorical violence". Am Anthropol 107(3):321–330

Skegg PDG (1975) Human corpses, medical specimens and the law of property. Anglo-Am Law Rev 4(4):412–424

Skegg PDG (1992) Medical uses of corpses and the "no property" rule. Med Sci Law 32(4):311–318

Skloot R (2010) The immortal life of Henrietta Lacks. Crown Publishing Group, New York

Steiner CB (2006) "En eller anden forbindelse": En etnografi om danske sæddonorer i et slægtsk-absperspektiv. Københavns Universitet, København, pp 1–130

Taylor JS (2005) Introduction. In: Taylor JS (ed) Personal autonomy: new essays on personal autonomy and its role in contemporary moral philosophy. Cambridge University Press, Cambridge

Thacker E (2005) The global genome: biotechnology, politics, and culture. The MIT Press, Cambridge

Thompson C (2005) Making parents: the ontological choreography of reproductive technologies. The MIT Press, Cambridge

Timmermans S (2006) Postmortem: how medical examiners explain suspicious deaths. Chicago University Press, Chicago

U.S. Congress, Office of Technology Assessment (1987) New development in biotechnology: ownership og human tissue and cells. Congress of the United States, Office of Technology Assessment, Washington, DC

Waldby C (2002) Stem cells, tissue cultures and the production of biovalue. Health 6(3):305–323

Weber M (1992) The protestant ethic and the spirit of capitalism. Routledge, London

Weiner A (1992) Inalienable possessions: the paradox of keeping-while-giving. University of California Press, Berkeley

Wilde O (2003) The picture of Dorian Gray. Barnes and Noble, New York

Witte J, Have HT (1997) Ownership of genetic material and information. Soc Sci Med 45(1):51–60

Wolfe C (2010) What is posthumanism? University of Minnesota Press, Minneapolis

Zelizer VA (1979) Morals and markets: the development of life insurance in the United States. Columbia University Press, New York

Zelizer VA (1998) The proliferation of social currencies. In: Callon M (ed) The laws of the market. Blackwell Publishers, Oxford, pp 58–68

Zelizer VA (2011) Economic lives: how culture shapes the economy. Princeton University Press, Princeton

Chapter 3
What Is a Human Body?

During the past couple of decades, the body has attracted massive attention in popular debate as well as in scholarly literature. We are living in "body times" to use an expression coined by ethnologists Susanne Lundin and Lynn Åkesson.[1] This booming interest in bodies is intriguing; as once remarked by Emily Martin, it indicates a period characterized by new types of challenges to bodies as we know them.[2] Nevertheless, notions about bodies often remain implicit in the commentaries on "new technologies of the body," "enhancement of the body," "sexualization of the body," and "(bio)medicalization of the body." Commenting on the surge of body literature, Janelle Taylor has pointed to "a tendency to presume, rather than ask, what a body is and where its significant boundaries are located."[3] She suggests, as have other scholars in STS, anthropology, and philosophy, that we need to begin to question how something becomes part of a body at a very basic level. This chapter addresses the call.

The bulk of the anthropological and sociological literature on bodies is characterized by a paradoxical double move. One the one hand, various poststructuralist approaches have dismantled bodies and notions of selfhood and subjectivity in search of ways of writing about human lives that defy universalizing categories. Partly this reflects a particular ontology of incessant emergence (and one that influences my thinking too); partly this ontology is tied up with a political project of defying the suppression supposedly purported by universalizing claims.[4] At the fore of this type of literature has been an interest in a fragmented, partial, and empty body related to a situated, constructed, and distributed self. On the other hand, an equally significant literature has focused on embodiment and phenomenological experience as a way of invigorating ethnographic inquiry without attributing explanatory power to abstract concepts like structure, culture, or society. This strain of literature tends to share the commitment to an ontology of emergence and to some degree even the political ambition serving as its undercurrent. Both sets of literature attempt to engage the world without reducing it to predefined categories and units of objective measurement (and potential oppression). But they approach the body in very different ways. Don Ihde has sought to bring these two approaches—the culturalized, socialized, and subjectified discursive body inspired by Foucault

K. Hoeyer, *Exchanging Human Bodily Material: Rethinking Bodies and Markets*, DOI 10.1007/978-94-007-5264-1_3, © Springer Science+Business Media Dordrecht 2013

and the phenomenologically lived body inspired by Merleau-Ponty—into dialogue in what is typically called postphenomenology. For Ihde, technological mediation of experience is the leading figure in this project.[5] This chapter similarly seeks to bring the two into dialogue, not so much through contemplation of technologies per se, but through the ubjects produced with technological assistance and in everyday practices.

In its own intriguing ways, medical usage of ubjects challenges notions of fixed boundaries of "the body" and presumed stable relations between such bodies and persons. As such, they instigate the type of basic questions alluded to by Taylor while simultaneously drawing attention to the immediate carnal experience of bodily space. Just as Bruno Latour suggests that hybrids precondition purification, I use this chapter to show how ubjects sustain a cultural imaginary of bodies as neat skin-bounded entities standing on a one-to-one relationship with persons of dignity and integrity. It is through new forms of transgressions that the boundaries, which ubjects are seen to transgress, become enacted in their current form.[6] Ubjects produced in medical practices, in particular in transplantation medicine, have proven themselves useful for exploration of new images of what it means to be a human being.[7] Challenges posed by transplantation medicine have not always led to intellectual invigoration, however, and it is not unusual for studies of organ transfers to ignore the Foucauldian, poststructuralist, feminist, and post-humanist literature on the fragmented body. Transplants are instead portrayed as challenging "fundamental" values of integrity and dignity.[8] These values are in turn linked to a body that presumably *used* to have clear boundaries. Thereby, the body is reinstalled as a special—enclosed—space, and techno-capitalism comes to figure as a novel intruder disturbing the peace. But peace never was complete. As argued by Linda Hogle, and most of the Latour followers, new technologies do not confuse natural body boundaries: they change the boundary-setting practice.[9]

To understand better how ubjects interact with practices of boundary setting, we need to confront prevailing ontological assumptions about bodies and their relationship to persons. The idea that transplantation medicine undermines integrity is central to many of the market opponents, as described in Chap. 1, and it typically involves a tacit understanding of bodies as persons. Indeed, capitalist medicine might very well undermine certain understandings of integrity for certain people, but we need a more specific understanding of why it is so than reference to presumably universal—natural—categories. Similarly, we saw how it is common for market proponents to make simplistic claims about body parts being "things." Such claims do not capture the subtleties in how we relate to and live with bodies and many of the ubjects they produce. I therefore ask what bodies are—and as indicated in Chap. 1, I answer this question with another, in my view better, question which is inspired by Annemarie Mol:[10] How are bodies *done*? As Annemarie Mol and John Law put it:

> ...the assumption that we *have* a coherent body or *are* a whole hides a lot of work. This is work someone has to *do*. You do not have, you are not, a body-that-hangs-together, naturally, all by itself. Keeping yourself *whole* is one of the tasks of life. It is not given but must be achieved, both beneath the skin and beyond, in practice.[11]

Rather than situating this work in political struggles alone, as some poststructuralists tend to do, it can be helpful to acknowledge phenomenological experience as part of the work going into the making of "a body." The phenomenological literature on embodiment—when not universalized into one fixed type of experience belonging to one transcendental body—can help us to include important dimensions of that work.

This chapter therefore first revisits the classical phenomenological literature and situates it in relation to reflections on subjectivity because ideas about subjecthood continue to influence controversies about ubjects. I then move onto literature which has challenged straightforward notions of bounded bodies corresponding to singular persons. Having dismantled the idea that persons equals subjects equals singular bodies, I begin a personal reading of skilful ethnographic studies of medical technologies involving the use of ubjects. This literature contains many important insights into the social life and attributes of ubjects and facilitates theorizing of what we might call ubjectology. I particularly wish to demonstrate ubjects' ambiguity, agency, and their complex affiliation with ideologies, politics, and conceptions of self and other. The work on these technologies also illustrates the problems of thinking about bodies as bounded entities. In the subsequent section, I confront these problems from a different angle by engaging the perspective from biology to show that there is no natural science to which we might turn for a naturally defined body. It is important to remember that ubjects are not solely products of new technologies, and therefore I turn in the subsequent section to classical studies of ubjects related in particular to death and defecation. This section also serves to remind the reader of classical insights into boundary making. Finally, in the last section, I acknowledge that by insisting on the notion of doing rather than being, I need to consider if there is anything that sustains the doing. Where is the room for *will* in this type of flat ontology of emergence that this book propagates? To avoid an implicit invention of a transcendental subject behind the doing, on the one hand, and to tie my discussion of the body to biopolitical struggles and strong interests, on the other hand, I end the chapter by bluntly confronting the issue of the will and where will is located.

Phenomenology, Postphenomenology, and the Issue of Subjectivity

The body is the phenomenological category par excellence. It is tempting also to say that it is a fundamentally meaningful category in as far as anybody who would engage in a discussion about bodies would also have "one." However, not all languages have a word for body, and not two bodies are alike. So what is it, this "body," that people are supposed to have?[12] It is not possible to list a set of characteristics that would define everything we (in various situations) would want to call a body— from the agile to the weak, from the unborn to the dead—and there are no clear points at which something becomes or stops being body. Nevertheless, most of the bodies that do in fact read a book like this would have a word for it (in as far as they would be able to read English). And the category of body is, not only in English, a sensuously and intimately meaningful word even if we may not be in a position to define its exact meaning. One of the aims of this book is to employ this quality *productively* rather than letting it do a lot of unacknowledged work for us. I therefore continue to use the word "body" and simultaneously engage the challenges that ubjects pose to any attempt of isolating one universal meaning. Indeed, I even use the intuitive understanding of the word body to define ubjects as that which leave the space identified as body.

Phenomenologists often say that a human being both *has* and *is* a body. The notion of *having* is ambiguous at best: What is it that can be said to "have"? Is it a mind? Certainly phenomenologists would oppose such a Cartesian bifurcation. And what does *having* imply anyway? The discussion of ownership metaphors in Chap. 1 made it clear that expressions like "my body" are very different from "my ball" and having a body is very different from having clothes to dress it. The notion of *being* is equally problematic as it does not solve the problem of defining what is and what is not (or no longer, at least) part of a body. Furthermore, the conjunction of being and having implicitly privileges the living and conscious body at the expense of the type of body found in the morgue or autopsy room from which many ubjects are taken.[13]

The well-known phenomenological expression is nevertheless a relevant starting point when rendering "the body" because it conveys a double status of bodies as meaning-producing subjects and as material objects—and Merleau-Ponty in fact used it to warn against reducing bodies to either things/nature/biology—as if they were just matters of materiality—or to intellectuality/meaning/hermeneutics—as if they were just clusters of semantic signs or producers of signs.[14] In a sense, this corresponds to the strong tradition in STS for emphasizing the coproduction of materiality and semantics.[15] There have been many more sophisticated philosophical discussions of what it means to be and have a body than what I just presented, but my point is to do away with the most dangerous reductions of the complexity it conveys (making "having" into ownership and "being" into a unit with clear boundaries) and instead emphasize the impossibility of definitively prioritizing materiality (and objecthood) over sociality (and subjecthood) or vice versa in one's accounts of "body." The emphasis on semantico-material coproduction is essential for understanding what ubject exchange involves, as this chapter will show, and it has become important for anthropologists too as they seek to move beyond the body proper as a "skin-bounded, rights-bearing, communicating, experience-collecting, biomechanical entity"[16] and instead explore the lived body as the domain of "neither a cultural mind nor biological body, but of a lively carnality suffused with words, images, senses, desires, and powers."[17] Merleau-Ponty insisted on the materiality of subjectivity but also on the inescapable interconnectedness between experiencing subjects.[18] Still, it remains a basic question—What is it that can be said to experience something? It is important to confront such ideas about subjectivity and selfhood because,

I will argue, they have important performative effects on notions of integrity and thereby on how ubjects are treated.

In a classical essay, psychologist Seymour Epstein seeks to mend the split between two strains of psychological literature that either takes the self for granted as the fundamental unit of study—without any attempts to define it—or disregards it altogether as unscientific and lacking in objectivity. Epstein suggests shifting the ontological claims from whether or not the "self" exists to a study of a pervasive empirical phenomenon: people having theories of a self.[19] I like this shift. We do not need to define subjectivity and selfhood as permanent and transcendental onto-logical entities. But we need to acknowledge that "theories" of subjectivity and selfhood are important because they feature as persistent ideas with significant biopolitical implications.

The granting of subjectivity is part of posing an actor as someone deserving respect, help, and support. This admirable aim typically leads to interventions which potentially involve reducing other actors to means for the *deserving* subject. History is full of examples of exploitation of people deprived of subjectivity and selfhood, and often this is done for the sake of helping somebody else deemed more worthy or deserving. Care is demanding; concern can be cruel. Even more often than peo-ple, animals have been deprived of subjectivity. Nichole Shukin argues that the involved anthropocentrism is characteristic even for much of the biopolitical litera-ture that otherwise claims to be critical of such processes.[20] This is despite the everyday experience that most people have of intersubjective communication with animals, which has been so vividly described by Donna Haraway in her book *When Species Meet*.[21] Subjectivity is a fragile quality, and understanding when and how it is attributed to various actors is part of understanding who comes to be seen as deserving care and what is seen as *just* a body.

To de-subjectify is to facilitate action. Some of these basic biopolitical mecha-nisms also do their work in relation to ubjects. When in a transplant procurement unit an ubject such as an organ or a piece of bone leaves the body, the first step of the disentanglement involves an element of de-subjectification; it must be trans-formed into materiality to begin its travels. For most people who agree to a dona-tion, this de-subjectification of the ubject is probably gratifying,[22] and for living donors one strategy can be to designate the ubject as waste, as I will describe in the next chapter. Successful de-subjectification is never certain though, which is of course why I insist on the notion of the ubject.

Experience of subjectivity and people's theories of selfhood inform what comes to count as a person in both a semantic and a material sense. Ubjects typically make a bad fit, in particular when people operate within the one-person-equals-one-bounded-body imaginary—what is sometimes called the *singleton model*. For many people the sin-gleton model seems to be part of their intuitive and embodied theory of self. However, many types of studies have challenged the universality of ideas about bodies as sin-gular skin-bounded entities relating on a one-to-one basis to persons as selves, and I will now introduce cases from the literature to allow reassessment of a mode of thinking that often informs opposition to "market"-like exchange forms.

Challenges to the Singleton Model

Anthropologists have long challenged the universality of the singleton model of complete material continuity between one body and one person. For some groups of people, personhood is supposed to follow the repeated use of a name in a clan so that a newborn acquiring the name of a recently deceased will be a continuation of the deceased person in a quite literal sense; for others, personhood moves back and forth between human and nonhuman animals, flowers, or trees; for yet others, it is exchanged between ancestral spirits and contemporary kin members.[23] The person/body link can apparently be narrated in an almost infinite number of ways. In a classical article about comparative moralities, Read suggested that the famous Cartesian split so prominent in the West made no sense in the context of the Gahuku-Gama where the person, according to Read at least, *is* his or her body in a very literal sense.[24] A phenomenological paradise, you might add, if not Read's analysis of Gahuku-Gama personhood ascribed a fundamentally distributed character to persons because you would literally share your person through touching, by drinking urine, etc., whereby persons run the risk of losing "their parts" to others too. Instead of rehearsing the full range of such studies, I wish to highlight just a few of them directly addressing the issue of transplantation medicine.

In her study of the Nayaka in Southern India, Birk David argues that the Nayaka basically does not experience the individual as a skin-bounded entity. If anything, skin is a point of fusion with others rather than parting. She therefore suggests that the typical Western criticism of bodily transgression in transplantation medicine does not make sense among the Nayaka.[25] Similarly, studies by Brad Weiss in East Africa and James Leach in Melanesia suggest that rumors about organ thefts are poorly understood if read through the lens of commoditization and infringement of bodily integrity: they relate to experiences of power, but in contexts with no obvious dichotomy between persons and commodities and with different perceptions of the body/person link, the transgressions involved in the supposed organ theft have nothing to do with markets undermining integrity. Organ theft rumors are ways of rendering experiences with technologies, with global and local inequality, and with violent relationships—but they are not stories about commoditization.[26]

Living Hybridity: From Cultural to Material Challenges

In general, however, there is a tendency in anthropological studies of the person-body link to overvalue culture (in the sense of meaning-making) and to overestimate cultural difference. Surely, the "meaning" of ubjects and their exchange varies, and what counts as a human body part does indeed vary immensely over time, institutional setting, and cultural and social context—but ubjects do something which cannot be reduced to cultural reading alone. And being-in-the-world and relating to bodies in all their plentiful manifestations have some aspects to it which go beyond cultural reading. Rather than seeing the one-person/one-body link undermined primarily by anthropological claims about other "cultures" understanding the body/ person link differently, there is something more immediately compelling over listening to people living lives that defy this sort of categorization in the most obvious sense, for example, conjoined twins. Alice Dreger shows in her research how conjoined twins can share the same body while having different spouses, dreams, and aspirations. She recounts how some who are not fond of food can enjoy the benefit of letting the other twin eat for both, but also the problems associated with coordinating your shopping if one person prefers using a list and the other prefers shopping on impulse. Having read Dreger's work, it seems absurd to hold on to ideas such as one-person-equals-one-body as a universal principle from which subsequent rules can be deducted.[27] The singleton model of absolute material continuity between body and person nevertheless remains the implicit assumption guiding a lot of the current legislation on tissue donation, which often defines tissue as "part of a person" and as falling under the same rules concerning informed consent as those applying to human subject research. In a lot of genetic research, this model is entrenched further through a reduction of body to "unique genetic code." Ironically, what is procured and stored in many genetic biobanks is of course material that is desired exactly because it is supposed to be shared among family members or fellow patients (e.g., DNA). Studies of microchimerism point out the specificities of this perception of one person, one body, one genome.[28] Recent microchimerism research has shown, for example, that at least 70% of all women, probably more, carry Y-chromosome cells (so-called male cells, though this gendering is obviously problematic) in their blood and thereby several "unique genetic makeups."[29] Again we are thrown into material hybridity in ways which ought to stimulate our thinking rather than just bolster the inclination to purification. Ubjects are entangled in fundamentally ambiguous relations with persons.

In the following, I move on to a personal reading of eminent studies of various medical technologies with the claim that many of them could in fact be coined as revolving around ubjects. These studies illustrate important characteristics of ubjects and testify to the usefulness of taking ubjects as entry points to exploration of bodily ambiguity. Furthermore, I think these studies and those presented in the following section invite us to think more broadly about life (in a biological sense) and lives (in a phenomenological sense) as processes of perpetual emergence; that is, I suggest they help us in understanding bodies as nodal points in perpetual semantico-material flows, not as static and well-delineated wholes encapsulating autonomous persons.

Doing Bodies in Medical Practices

When Caroline Walker Bynum suggested that the singleton model of material continuity frames many of the public controversies surrounding tissue collections and transplantation medicine, I think she overlooked the need to attribute some element of agency to the transplanted ubjects and not just the surrounding culture. Ubjects hold a *potential* that can be unleashed irrespective of the predominant culture (here understood as dominant discursive practice). In the USA, for example, transplantation medicine has had strong institutional backing in its attempts of objectifying the transplanted ubjects. Nevertheless, there is something uncanny about organs; some element of affiliation with their origination seems to persevere. Important work by Lesley Sharp and Margaret Lock, for example, has shown how relatives to organ donors and recipients defy the dominant medical attempts of objectifying the transplanted organs—the ubjects.[30] These people often make their own interpretations of what organs bring along besides the biological functions that instigate their transfers in the first place. Some recipients experience a new craving for chocolate that they attribute to the organ or they increase the amount of sport that they undertake to satisfy what they perceive to be the needs of a heart from an athlete.[31] Strong emotional ties can emerge from these transfers when relatives and recipients find one another through Internet forums or support groups. Based on studies of living organ donation, Aslihan Sanal notes that such ties can also be experienced as most uncomfortable when a donor feels entitled to interfere with how the recipient "treats the organ."[32] Lock and Sharp show how also medical professionals in fact relate ambivalently to many ubjects and, for example, avert from transplanting the heart of murderers or feel hesitant about brain death. We could say, of course, that a deeper level of culture (identified by Bynum) overrules the institutional efforts of creating a dominant culture of objectification. I think, however, that it is more fruitful to direct our attention to the agency of the ubject and the phenomenological experiences it provokes.

In my own work on bone transplants, it appears that most people can donate and receive bone without considering issues related to material continuity in any depth.[33] And it goes for many other types of simple, mundane, routine transplants that they have never instigated ethical inquiry or popular concern.[34] It is not every ubject that causes concern. When, however, an implant begins to demand work of recipients, if it causes pain, for example, recipients begin to think about its origin. In one case in my own work, a young woman's knee would not heal, and she began contemplating the implanted bone in new ways; it became alien and threatening and made her go for checkups. Even though she was reassured by the doctors that it was not an issue of tissue compatibility, she kept contemplating the origin of the bone as a potential reason for the pain. Organ recipients have to do some work every day to retain the organ's functions. I think this work (and not just a cultural model of material continuity) produces reflections. Ubjects are never fully under control. As socio-material entities, they exert wanted and unwanted agency: they can infect, transmit diseases, cause pain, worry, inspire, provoke, scare, and sometimes save lives. They overflow. Studies of ubjects therefore involve an analytical possibility of overcoming assumptions about coherent and monolithic cultures attributing meaning to a passively awaiting nature.[35]

Potentialized Subjecthood

We can develop our sense of the potential agency of ubjects further if we relate it to Marilyn Strathern's thoughts about partial connections.[36] Through partial connections, part and whole can change place: it is unclear what stands in for what. Are gametes persons or part of persons; part of whom (male donor, female recipient, or potential future child); and by way of whose agency (the donor, the recipient, the gamete, or the procurement organization)? It depends on the view taken, the connections made, and the action through which its exchange unfolds. In the words of Marilyn Strathern, parts "are not fragments of whole unitary identities; if they are parts of anything then we might say that they are parts of relations."[37] The partial connections that can be drawn are without given limits. In my view this complexity makes it naïve to presume that at some point *everybody* involved in such transactions will subscribe to the view that the involved cells are "just objects," and objects of a particular "kind," to be traded between producers and consumers. The partial connections that can be made interact with vital institutions—such as kinship—and they are inherently difficult to control.

The precondition for most contemporary medico-technical developments is the technological facilitation of *in vitro* life, biological activity of tissues and cells outside the carnal chamber loosely called body. In her work tracing these developments, Hannah Landecker suggests that cell cultures have become sites through which the plasticity and temporality of life has become radically rethought.[38] The life in the petri dish has facilitated the proliferation of new biological mixtures that concerned ethicists often lament. The *in vitro* research frontier of tissue engineering, stem cell research, and xenotransplantion has attracted a lot of attention and stimulated important insights into the making of boundaries through perceived transgressions.[39] Life outside bodies is paradigmatic for the development of the ubjects attracting popular attention and concern, but in fact most of this laboratory work has never attracted much attention.[40] Even one of the most notorious examples, the HeLa cancer cell line used in research laboratories all over the world, was not seen as an ethical problem at the time of its production. HeLa was generated from cells taken from a black woman by the name of Henrietta Lacks (hence the name of the line) who died in 1951. It became widely used, and tons of it, quite literally, are said to be in current use. In a recent bestseller about the intertwined history of HeLa and the descendents of Henrietta Lacks, science journalist Rebecca Skloot refers to the cells *as* the person Henrietta Lacks. Whether there is more of Henrietta Lacks today than at the point of her death, however, depends on your interpretation of the ubject—What is actually the relationship between a cell line and the person in whom the first cells originate? A cell line arises out of complex assemblages of feeder layers, scientists, laboratory staff, and assiduous cells. Its genesis is poorly understood if the agency involved is reduced to a genetic code of a cancerous cell.[41] Furthermore, there is a tragic irony to the claim this code *is* Henrietta Lacks, considering the role of the cancerous cells in killing the woman. I prefer thinking of HeLa not as a woman but as an ubject, the ambiguity of which is emblematic for many other ubjects.

Even if in vitro life has a sort of high-tech research frontier ring to it, it consists mostly of routine work and revolves around ubjects few people care about. The fact that HeLa existed for years before becoming an "ethical" problem is illustrative and points to an essential element of the argument that this book is making. Most ubjects live very quiet lives. But they embody—because of their disembodiment, that is, their relationship to bodies and through them to subjects—a potential for future conflict. The taking of blood samples in medical settings is typically viewed as so utterly ordinary and mundane that Pfeffer and Laws found it very hard to engage anybody in a study of it.[42] Nevertheless, the massive legal and ethical attention to biobanks during the past decade illustrates how the very same samples that cannot interest anybody at the point of procurement can cause substantial institutional reactions when used in new institutional settings or for new purposes.[43] Ubjects are unruly. As a consequence of this unruliness, I suggest that it is unwise to build public policies on the view that body parts are just objects. Conversely, it is also misleading to attribute a stable, inalienable, and inert quality to all blood samples, cell cultures, organs, gametes, or other ubjects. Ubjects are neither things nor persons, but they do have a *potential* for making unruly partial connections.

Sociopolitical Context and the Inability to Bring Closure

One of the most industrialized, routinized, and widespread assembly lines for producing ubjects is the blood product industry. Through the work of Titmuss (see Chap. 1), and later Paul Rabinow, blood donation has also delivered theoretical models for debates about procurement of other ubjects.[44] What Titmuss highlighted was not only national differences between the USA and the UK, but how the "altruism" needed to produce ubjects without compensating donors is institutionally entrenched. In more recent work, Kieran Healy has pursued this approach in comparison of different blood donation systems. He has shown how donation patterns reflect institutional contexts and how altruism can be analyzed as a social product.[45] Jacob Copeman along the same lines points out how anonymous, voluntary, and unpaid donation can be very different things, depending on institutional and cultural context in his comparison of British and Indian blood donation systems. In India, gurus are supporting a shift to nonpaid donations and imbue voluntary donations with religious meaning.[46] Anonymity in donation practices does not necessarily imply alienation, he explains; it can facilitate establishment of other relations to, for example, religious or national communities.[47]

Just like reproductive, transplant, transfusion, and cell-culture technologies, device and implant technologies continuously produce new types of ubjects. As already mentioned, the literature on tissue economies rarely deals with such nonbiological ubjects. However, these ubjects also provide fascinating arenas for challenging the notion of the person as a skin-bounded entity—and they also exert a lot of unwanted agency. Even the most mundane and well-entrenched technologies can stimulate reflections about what is part of body and self, when they involve moving ubjects in and out of bodies. Devices interact with gender, national politics, and class inequalities.[48] Many devices have roots in veteran rehabilitation programs. David Serlin has argued that "normalizing" veterans was part of a gendered strategy of covering up the implications of war in a strongly homophobic USA that could not cope with passive and care-demanding men.[49] Julie Kent and Alex Faulkner have pointed to the gendered regulation of implant technologies in the EU through comparing governance of hip and breast implants.[50] There is a long history of disrespect for women undergoing breast operations, from the 1920s when the techniques were first used for reducing breast size as part of the general application of plastic surgery techniques (developed as a sad sequel to World War I) to the many subsequent attempts of enlarging breast with fat, paraffin, or silicone. A persistent trait of these debates has been reflections on the agency involved. Women have been portrayed as acting against their own interests—as socioculturally determined and suppressed.[51] Some women, however, claim to experience a boost to their sense of self thanks to successful implants—So why relegate agency to suppressive structures and represent women as incapable of choice?[52] Importantly, Julie Kent also points to the agency of the implant. The implant does something to the sense of self whether for the good or the bad. Sometimes the silicone leaks; for example, it becomes the cause of pain and then women describe it as an "alien threat." At this point it stops being part of them and works against their sense of agency.[53] A similar move from being "part of me" to "alien threat" was described above in relation to bone transplants when they began causing pain. Haraway has suggested taking our ability to melt together with technological devices seriously and to consider ourselves cyborgs.[54] But being cyborg is not always easy, and the determination of what is part of whom (or what) reflects the agency of the implanted ubject as well as the people making the judgment.

All these studies show how ubject exchange interacts with understandings of bodies and relates to political, economic, and moral agendas. They point to the open-endedness of such processes. I think they thereby also illustrate the audacity of the many regulatory attempts of stabilizing and hegemonizing particular understandings of the body proper. Bodies—as we get to know and identify with them— emerge, change, and effuse clear identification; they have no clear beginning or end, not in time nor in space. This basic ambiguity implies also that what the ubjects traversing this space *are* and what they do are poorly accounted for if we *define* them as either "things" or "bodies." The inability to pinpoint body boundaries becomes even more obvious when I now turn to studies of immunology, bacteriology, and genetics—research fields that deal directly with understanding and conceptualizing bodies. As with Dreger's study of conjoined twins, we can use exploration of biological variance in these fields to provoke our thinking about bodies.[55]

Bodies from the Perspective of Biology

It is not always obvious to the biological and medical scientists themselves that their attempts of categorization do indeed challenge the notion of a skin-bounded entity. In their studies of immunology, Emily Martin and Donna Haraway have elucidated how immunology as a discipline has been as much about constructing bodies as units (using metaphors of the nation state in need of armed defense) as about exploring the actual relatedness of life forms that take place in the studied bodies. This work has been followed by more recent biological inquiries into bacteriology such as those converging in the NIH-initiated Human Microbiome Project.[56] Presented as a sequence to the Human Genome Project (HGP), this new attempt of mapping the different organisms that occupy bodies in many ways departs from the conventional HGP idea of the body as a product of genetic and environmental interaction because the environment becomes life within. This shift is central for what I want to highlight in conjunction with ubjects: notions of inside and outside, part and whole, me and not-me are coproduced with different ways of exploring the body. Initial findings from the Human Microbiome Project indicate that different spots on the body constitute different habitats, but also that these habitats change in cycles over the course of a day as well as gradually throughout a body's life. People can have very different mixtures of microbes, and certain mixtures might be of decisive importance in, for example, development of obesity.[57] Bacteria can be friends or foes, us or alien, depending on what persons *want* to happen.

For decades, people like evolutionary biologist Lynn Margulis have used bacteriology to propagate a different understanding of bodies, not as stable units, but as communities that have emerged through complex coevolution.[58] She has pointed out how a body's dry weight usually consists of 10% bacteria and how many of the most basic metabolic functions are performed by organisms that do not share the human genome. Following a study of Margulis' laboratory, sociologist Mary Hird sums up how "we" consist of

> …600 species of bacteria in our mouths and 400 species of bacteria in our guts, and … countless more bacteria that inhabit our eyes, anuses, and skin. Indeed, the number of bacteria in our mouths is comparable to the number of human beings that have ever lived on earth. The number of microbes in our bodies exceeds the number of cells in our bodies by 100 fold. The human distal gut contains more than 100 times as many genes as our human genome (which has 2.85 billion base pairs).[59]

Bacteriology became Hird's entry point to rethink ideas about body boundaries. Biology is about coexistence and exchange; it is about gifting as she puts it, but as anthropologists have long known, gifting can be violent. Some encounters between people do indeed involve deadly exchanges. There is no teleological purpose with coexistence, and living implies no given units, no clear boundaries, within which objectives can be located as naturalized purposes of life. Just as Mol, Haraway suggests that boundaries result from work: "Bodies as objects of knowledge are material-semiotic generative nodes. Their boundaries materialize in social interaction; 'objects' like bodies do not preexist as such."[60] In a classical essay in theoretical biology, Stuart Kaufmann points to the hopelessness of trying to define a biological system based on biology alone, because it would presuppose a purpose that cannot be deducted from biological data: "An organism may be seen as doing indefinitely many things, and may be decomposed into parts and processes in indefinitely many ways … there is no uniquely correct view about what an organism is doing."[61] Later,

I wish to return to this point in the company of Canguilhem and his thoughts about distinguishing the normal from the pathological, but for now the point is simply that there is no biological level of truth to which we can turn in an attempt of identifying bodies as isolated bounded units.

Nevertheless, biological research narratives continue to deliver popular images of the essence of a human being. From fingerprinting to DNA typing, these images furthermore tend to interact with purposes of surveillance and control of the "units" identified with biomedical means.[62] In her book *The Century of the Gene*, Evelyn Fox Keller pointed to genetics as one such image that captured the popular and scientific imagination and provided a frame for identification of an individual's supposed biological essence. Genetics coincided with a series of interdependent technoscientific developments—not least with respect to information technologies—that supported what has been called an *informatization* of the body.[63] Bodies were increasingly reduced to a code, stable and fixed, which could be accumulated, processed, transmitted, and patented, as pointed out in Chap. 2. But following the completion of the Human Genome Project and with subsequent developments in systems biology, epigenetics, proteomics, and microbiomics, this rendering of a stable essence has proven to be as futile as other attempts of isolating a primary and stable essence that might capture what a human being is.[64] Or as productive, we might say, because even if genetics has not managed to set itself as the final answer, it has indeed produced many new entry points for intervention into the type of living we associate with bodies and served to produce massive collections of ubjects—the so-called biobanks—for research and surveillance purposes. But as scientific explanations, the initial prospects of a stable code delivering the blueprint of humanness have floundered. Genes are not the causal agents coding for everything else; they need to be read by something, and to understand this process of reading (if we accept the informatized metaphor), scientists are hurled back into the incessant exchanges of biological life.

The study of genetic variation in, for example, the HapMap Project implied according to geneticist Paul Dear that by 2004 we had to give up on the idea of one human genome. In 2009, the implications of research into copy number variations (CNV) within a given individual (which refers to small variations between the genome in various cells—the phenomenon which is most clearly expressed in cancer cells) led Dear to suggest that we should also write the obituary of the *individual* genome. We would need whole-body biopsies to determine such a genome, he asserted.[65] And what should we then say once we consider microchimerism?! Bodies are not easily captured as biological units; they are part of interdependent biological processes with no given boundaries or purpose, no obvious beginning or end. And in fact this basic premise has long stimulated interesting work in philosophy and anthropology. In the following I turn to classical studies of the inescapable flows that bodies engage in as they defecate, transpire, get sick, and die to tease out how insights from these studies can be fruitful in understanding our current handling of new ubject types in biomedicine.

Revisiting Classics and Classical Blind Spots

Living bodies defecate, urinate, transpire, and bleed (some even in cycles from puberty to menopause); they shed tears; their noses drip; they usually grow hair and nails that must be disposed of; and they produce saliva and other liquids rarely mentioned in decent academic texts. To delineate themselves from the material world they are part of, living bodies must perform various kinds of work. At some point, all bodies are destined to become part of material flows related to a process called *death*. Julia Kristeva portrays living as a feeble scuffle through which you try to purify yourself and lift yourself above that materiality onto which you are certain to fall prey:

> These body fluids, this defilement, this shit are what life withstands, hardly and with difficulty, on the part of death. There, I am at the border of my condition as a living being. My body extricates itself, as being alive, from that border. Such wastes drop so that I might live, until, from loss to loss, nothing remains in me and my entire body falls beyond the limit—cadere, cadaver. If dung signifies the other side of the border, the place where I am not and which permits me to be, the corpse, the most sickening of wastes, is a border that has encroached upon everything.[66]

Dying is a messy thing, as James Green notes in his book about "modern dying," and yet there is no such thing as a natural death.[67] Dying is imbued with meaning, and it is a process that can take an infinite number of shapes. Since we cannot turn to biology for definitions of biological systems, we cannot expect biology to determine when a system is dead either. In an era of transplantation medicine, this has important implications for the production of ubjects. Organ transplantation involves seeing the material body as composed of living entities, which can be used when the "person" is dead, a bifurcation of the subject and the object. This bifurcation at the point of death might not be totally alien to Merleau-Ponty who wrote that "If the patient no longer exists as a consciousness, he must then exist as thing,"[68] but I think the transition to thingness is not quite so smooth. As pointed out by Anja Marie Bornø Jensen based on her studies of donor relatives and clinical practice, organs are treated with an ambivalence entailing both subjecthood and objecthood.[69] We might say that in the shape of the cadaver the body as such turns into an ubject arousing uncanny and ambiguous feelings.

Death, Dirt, and Purification

Death is of course also a classical anthropological topic which has recently seen a surge of renewed interest.[70] Bodies do not just "fall beyond the limit" (to borrow Kristeva's phrase); usually a corpse is carried across. In anthropology there used to be an elaborate tradition for studying death rituals and also for relating these rituals to other rituals surrounding ubjects such as nail parings, hair, feces, urine, sweat, and blood. In his classic book on religion, *The Golden Bough*, James Frazer wrote with a touch of cocksureness:

> The notion that a man may be bewitched by means of the clippings of his hair, the parings of his nails, or any other severed portion of his person is almost world-wide, and attested by

evidence too ample, too familiar, and too tedious in its uniformity to be here analysed at length. The general idea on which the superstition rests is that of the sympathetic connexion supposed to persist between a person and everything that has once been part of his body or in any way closely related to him.[71]

There is an irony in Frazer feeling personally superior to such "superstition" while still suggesting universal relevance of his proposition. Cannibalistic practices such as the eating of the genitals of a deceased spouse or religious uses of ubjects as fetishes used to provide wonderful food for thought for anthropologists (if I may be excused the pun).[72] The handling of mundane bodily material was later discussed by Leach and Douglas as prime examples of the relationship between public and private symbolisms. Their work gave rise to significant insights into issues of categorization and how persons use bodies and body products to exhibit particular social relationships and commitments.[73] Basic and mundane fluids such as spittle, blood, and tears are powerful means of communication: most people would probably agree that having another person's spittle in your face is problematic for reasons other than the material-biological health hazards it might involve.[74] As remarked already by Durkheim in his classical work on religion (to which I will return in Chap. 5), body fluids and in particular blood are often used in rituals to consecrate sacred objects.[75] Ubjects possess great symbolic power.

Much like the ubjects studied by Leach and Douglas, the new therapeutic tools of transplantation medicine defy easy categorization as either part of or not part of human bodies, as do the implanted devices discussed above. They are also powerful symbols. Despite all these reasons to revisit the anthropological canon concerning handlings of cadavers, body parts, and parings, the legacy from the anthropology of religion seems to have fallen into disrespect, almost as if anthropologists and other social scientists unwittingly adopt Frazer's implicit stance: rituals and belief systems set up to create purity in relation to ambiguous bodily material express a type of superstition alien to the rational actors of contemporary biomedical institutions. In my view, we should refrain from such naïve assumptions about the societies in which we live and instead consider what we might learn from the classical studies.

In her seminal work on categorization, Mary Douglas noted how "all margins are dangerous" and how bodily parings simply by traversing "the boundary of the body" become potentially dangerous.[76] As matter out of place, they destabilize categorizations of immense symbolic importance. Drawing on Douglas (and van Gennep), Victor Turner explored the role of rituals in handling the categorically unclean, the liminal. He argued that rituals instigate a "change in being" and that ambiguity is central to such transformations. Turner described the position betwixt and between dominant categorizations as a "fruitful darkness" seething with a potential for cultural creativity.[77] The section above on ubjects in medical practices is an illustration of what a fruitful darkness can do. Susan Squier has criticized the applicability of Turner's understanding of liminality for an exploration of the new biological products of biomedicine.[78] Her point is that Turner talked about liminality as a purely cultural phenomenon, a question of meaning applied to an amorphous but unaffected nature. He argued that the human mind needs distinctions that biology cannot accommodate; hence, culture creates boundaries to make nature comprehensible. Also in Douglas' work there is a nature/culture distinction (in her case between dirt and defilement) whereby categorization becomes a purely social activity. Materiality is thus construed as the passive victim of culture, as it were, Squier points out. It is an inadequate description of ubjects. Still, the productivity of the undefined in the space betwixt and between is instructive, and it appears to me that Douglas' work on the dynamics of categorization serves as an often unacknowledged undercurrent to what STS scholars today mostly talk about as processes of hybridization and purification.

In linguistics, too, a lot of work has explored categorizations of the body. Some scholars have looked into the historical shifts in medical categorizations of body parts; others have sought to explore cross-cultural principles. What becomes clear from these exercises is that there is no abstract body out there that can supply categories of thought independent of the bodies interacting with and thinking about them.[79] Rather than searching for the right categories, it appears more fruitful to explore what categorizing produces and in this exploration appreciate the effects of the ambiguous. And if we abide to the reading I made of Kristiva above, we should see the act of categorizing what is part of a human being as constitutive for and deeply interlinked with social and biological reproduction.

In *The Civilizing Process*, Norbert Elias showed how a study of how we deal with bodily secretions can be used as a lens to social change. There are separate chapters on changing attitudes to blowing the nose, defecation, and spitting. His argument is that our attempts to control our bodies reflect social relations through which we construe ourselves as persons (and agents of power).[80] Less meticulously, but perhaps more radically and ferociously, Dominique Laporte once suggested rewriting the history of subjectivity as the *History of Shit*. Much like Kristeva, Laporte suggested that "doing" a subject is as much about undoing, getting rid of that which is not you, vested cleansing of waste, and dedicated purgation. He took point of departure in a royal edict from 1539 that made the individual Parisian household responsible for its own feces, and the book renders the handling of shit central for an understanding of the relationship between individual and state.[81] Indeed, there is a complementary sequel to the historical processes outlined in the previous chapter which would trace the corresponding changes in ways of *doing* one's body. Such an intimate history would need to include also how changes in private practices, the bodily technologies of self, are reflected in architectural changes in bedrooms, kitchens, and bathrooms, as seen in Elizabeth Cromley's work.[82] Basic acts of purification are ingrained into material structures in a very direct manner in the shaping of architectural design. An intimate history of this kind could, potentially, also introduce a much needed phenomenological dimension into the account presented in the previous chapter.

Superstition as a (Problematic) Analytical Category

I believe the everyday handling of body boundaries is related to the dynamics involved in the handling of the ubjects produced as part of new medical technologies. Ubjects are part of our lifeworld. Therefore, contemplating our daily material flows—and all the other instances of bodily flows that make up human beings as we know them—in light of the insights from Serres and Latour as well as Kristeva, Elias, Douglas, and Turner might help us understand the depth of the institutions involved in ubject exchange. And I wish to suggest that *it might also help us overcome the superstition that ubjects provoke just the superstitious.* Relations between bodies and persons can apparently be made in an infinite number of ways, but everybody needs to do some amount of work to figure out what they consider part and not part of themselves. The anthropologists suggesting that "their" people see no link between materiality and personhood whatsoever might be as naïve about the potentiality of ubjects as the biomedical procurement agencies who think donors and recipients soon learn that organs, tissue, and other ubjects are really "just" objects. No public "educational" campaign can put ubjects to permanent rest and settle the status of organs, tissue, or bone as plain objects so that they may be treated like uncontroversial commodities (or semisacred gifts). Ubjects are tied up with very intimate and enduring practices that reflect living as incessant semantico-material flows bound up with experiences of subjectivity.

Rendering of the inescapability of the everyday mundane semantico-material flows forces us to come up with something better than "intrusion of bodily integrity" when explaining what it is about some ubjects that make them so controversial. Bodies are entangled in flows, yes, but not all of them provoke or challenge institutions and phenomenological experiences of self. At some point, all bodies "fall beyond the limit—cadere, cadaver" and become materiality with no affiliation with persons. It is the interesting time *before* this happens that I wish to explore with ubjects, and the ontological claims about ubjects as an unavoidable feature of human life (so unlike what is currently in fashion in anthropology and STS) are meant only as a starting point for exploring the endless variance in what this ambiguity is producing.

If the history of shit is relevant for an understanding of the fundamental inability to distinguish person from body and body from materiality, it is also relevant for the return of the notion of subjectivity. In the next section I make a precarious attempt at contemplating what draws bodies out of materiality and animates them in ways that make bodies *matter* so much to most of us.

Will, Pathology, and Postcapitalist Medical Practice

Even if the problem of body boundaries can be solved by saying that making a whole of a body demands a lot of daily tasks, some of which are considered private and uncontroversial (but all of which are socially ingrained), it still seems too easy to say simply that the body is done through work. What keeps this work going? Facing this question is dangerous terrain for the poststructuralist because unless one is willing to talk in only the most abstract terms of culture, power, or discourse, this touches on the notion of *will.* I tried to do away with the ontological premises of the "I" in the sense of a transcendental self by way of installing embodied theories of self as an empirical phenomenon. It is tempting to do the same with will, but I would rather face it bluntly and suggest that we do need a more substantial concept of will to understand ubject exchange—at least if we wish to take the biopolitical framework seriously and acknowledge that there is a lot of craving for ubjects. We also need a concept of will unless we wish to subscribe to a flat ontology of material determinism of the type occasionally propagated by the logical positivists.[83] A useful concept of will should not hinge on a transcendental self, and it should not be too abstract and disembodied. It must be a semantico-material concept of will that can embrace both the poststructuralist and the phenomenological traditions. The philosophers of will who can provide a framework for this are Arthur Schopenhauer and Friedrich Nietzsche. They inspired not only Foucault but also the existentialist philosopher Søren Kierkegaard and through him phenomenologists such as Merleau-Ponty.[84]

The notion of will is foundational for Schopenhauer's philosophy. He challenged Kantian epistemology by discussing knowledge as a product of *interests*, and he made basic bodily cravings including sexual desire a legitimate topic for continental philosophy (well, more or less legitimate). For Schopenhauer the will is beyond explanation and without cause. It is not a person's property, so to speak, and not under his or her control. Control of will would presume some sort of hidden self choosing the will and, thereby, a regress into multiple selves behind selves.[85] According to Schopenhauer, it is meaningless to talk about the will as a cause of other things; in fact, philosophy should in his opinion refrain from inventing that type of *causa sui*. With Schopenhauer, there is no way of knowing the will before it manifests itself in doing. Doing is everything. You do not know your will until you are confronted with something obstructing your doing. Pain and suffering can provide such moments of self-realization. Such obstructions are *productive*; they give a sense of direction. Freud is said to be inspired by Schopenhauer in his theory of the unconscious, but for Schopenhauer there is no obvious therapeutic solution. He was inspired by Buddhism and the notion of liberation from worldly longing, but he basically saw human beings—bipeds as he derogatively called us—as caught in between states of oblivious boredom (for those who do not realize their will due to lack of obstruction) and suffering (due to overwhelming opposition).

Nietzsche took a more optimistic stance in his appropriation of Schopenhauer—he celebrated the will! He contends that we need not know where it comes from, but we should enjoy its fruits. Nietzsche notes that "in real life it is only a matter of *strong* and *weak* wills," but like Schopenhauer, he refrains from inventing some sort of agent behind the will.[86] He strongly opposed the Kantian self-restraining attitude that construes a subject choosing every act: "There is no such substratum; there is no 'being' behind the doing, effecting, becoming; 'the doer' is merely a fiction added to the deed—the deed is everything."[87] The doer added to the deed is, nevertheless, a persistent trait in the Western bioethics tradition with its emphasis on autonomy. But Nietzsche showed little respect for it:

> The desire for 'freedom of the will' in the superlative metaphysical sense, which still holds sway … involves nothing less than to be precisely this *causa sui* and, with more than Münchhausen's audacity, to pull oneself up into existence by the hair, out of the swamps of nothingness.[88]

Will is not a choice, but it *is* foundational for human activity, and unlike many of the popular uses of Kant, Nietzsche does not rebuke the person inclined to pursue an inclination.

Will and the Sense of Control

If the will manifests itself through confrontation with pain and suffering, the phenomenologist will emphasize that somebody experiences the pain and undergoes the suffering. It is not a transcendental subject behind the act, but somebody engaged in an all too material being-in-the-world. Parallel to Schopenhauer's thoughts about absence of will until the point of confrontation and opposition, Drew Leder from a phenomenological perspective describes the body as absent until confronted with something. Persons usually do not feel their bodies; they feel with them. Then, especially when confronted with illness and pain, the body demands attention. It is objectified. The similarity with Schopenhauer's point is obvious, though rarely alluded to. Suffering creates awareness of the body as a material object, and it creates yearning. Michael Jackson suggests contemplating this potential for objectification of one's body in light of the sense of *control*.[89] That which is experienced as sustaining the sense of control is usually deemed part of me, while that which is threatening or seemingly out of control is deemed not-me. Narration can be a mode of regaining a sense of control in confrontation with confusion and suffering, Jackson argues,[90] and yet Schopenhauer advises us not to look for a subject hidden behind the urge to narrate. Will is performed through narration; it does not cause it. Many of the examples given above of implants and devices either helping or hampering persons wanting to achieve particular agendas can be understood as instances of ubjects being defined as either me or not-me, depending on the phenomenological sense of control. Implants and devices become alien threats when they hurt. Tissue can also become viewed as alien. Cancerous cells, arthritic hips, and bacteria can through technological mediation and scientific classification become identified as not-me, again in reflection of experiences of lacking control.[91] Bodies have no boundaries until we draw them, and today such drawings often involve complex biomedical networks. But not solely networks: they demand willing.

I wish to draw this connection between a phenomenological notion of bodily control and a Schopenhauerian sense of will a little further by relating it to Canguilhem's rendering of the normal and the pathological. My aim is to move more definitely beyond the recurring culture/nature divide and locate will in a socially ingrained yet thoroughly biological body with a propensity for casting theories of self. Georges Canguilhem's two seminal essays published together as *On the Normal and the Pathological* avoid locating normality and pathology in either biology or sociality.[92] In line with the argument above, Canguilhem points out how there are no given boundaries for a living system and no given purpose for living. Cancerous cells are as living as any other cell in a body even if they can be classified as pathological: "…it is medically incorrect to speak of diseased organs, diseased tissues, diseased cells. Disease is behavior of negative value for a concrete individual living being in a relation of polarized activity with his environment."[93] People are not indifferent to the living taking place, and when conscious and demanding beings experience and interpret living, their longings are part of shaping it. Canguilhem talks about pain as experienced by a "conscious totality" (thereby giving us yet a concept for the dangerous terrain of subjectivity discussed by phenomenologists and philosophers of will), but this totality is not just a biological sensory mechanism. People tend to evaluate even pain through norms that they cannot decide upon independently of the societies in which they live. In short, there is no way of giving primacy to either social norms or biological manifestation: "There is no fact which is normal or pathological in itself."[94] As norms change, the definition of the pathological also changes. To emphasize the perpetuity of this process, Canguilhem preferred talking about normativity rather than normality, the former being cast as an activity rather than a state. Through normativity, persons learn to rinse themselves of that which is not proper. Getting to know what to rinse out (in the sense of Laporte) demands knowledge, and this knowledge is always socially constructed and always charged by interests.

Socio-material Will

The point is that a body does not just want a new liver, kidney, or heart. It is not a primordial longing. And yet no disembodied will exists; longings and cravings are more than "social constructions." Socially entrenched persons learn to interpret obstructions to their will as something to be alleviated with a new organ, a piece of tissue, or more blood. Every will struggles with other wills—there are strong wills and weak wills as Nietzsche says. We might add that different wills are supported unequally by various societal institutions. If organs can be viewed as property objects, the market institution delivers a much appreciated support for people of wealth for their longing for a new kidney; for people of fewer means, it is not quite as appealing.

Like Schopenhauer, Canguilhem warns against believing in an easy fix or final answer to satisfy people's cravings. Norms change with every intervention, but the processes of biological normativity do not end, and "…to dream of absolute remedies is often to dream of remedies which are worse than the ill."[95] The question is whether some cravings for ubjects are instances of such remedies producing something worse than the ills they seek to alleviate. The current practice of medicine in some respects reflects what Melinda Cooper calls a postcapitalist delirium characterized by a utopian pursuit of perpetual growth and boundless health. There is no limit to what you should aspire for. Better than well, as Carl Elliott frames the motto of American medicine.[96] The postcapitalist logic recognizes only growth, no Aristotelian balance or middle way. This logic currently seems to be interacting with the vital game of normativity described by Canguilhem, whereby it stimulates a voracious appetite for ubjects. The boundless body is drawn into a boundless pursuit of more health, but not everyone has the same means to manifest their will in the ensuing power struggle.

Making Wholes

There is no way to define a body, no biological level onto which we may turn for an answer to what a body is, and no cultural agreement on what bodies mean and how they relate to persons. Commenting on the phenomenological expression with which I began this chapter, Mol and Law remark: "We all *have* and *are* a body. But there is a way out of this dichotomous twosome. As part of our daily practices, *we also do (our) bodies*. In practice we enact them."[97] This chapter has endeavored to ask what keeps the doing going and why. It has also explored some of the practices through which bodies are done. These practices are biological and social. They have phenomenological dimensions and they reflect political contexts. "Body" does not refer to an entity; it is a word we have learned through immersion into language games, and these games keep changing in tandem with changes in the way we do bodies.

With language games, I of course paraphrase Ludwig Wittgenstein's late work on meaning as emerging in practices.[98] Language games define what is and is not part of a body as they evolve in interplay between epistemology, morality, and power. This takes us back to the wider theoretical framework of biopolitics. With this chapter I have sought to pay more attention to phenomenological and biological aspects of this interplay than is usually the case in the biopolitics literature. I have tried to take biology seriously but also to show that it does not deliver a foundation from which values can be derived. It does not tell us what a body is. Body boundaries are coproduced with subjectivity, and both demand work. I have gone further down this dangerous (and in anthropology and STS quite unpopular) road and contemplated will as that which keeps the biopolitical machinery going. This will relates to experiences of subjectivity, and it constitutes a source of care that can have perilous effects for those with few means of defense.

The will is important because there is phenomenological density in the basic empathic experience of confrontation with persons wanting something. This sense of will is what lifts a person from pure materiality in Kristeva's sense. We get inspired by will; we get frightened by it. Actor-network theorists insist on a principle of symmetry between human and nonhuman actors. For some, this principle has been used to erase will as an analytical concern or to multiply it so that objects and subjects have been ascribed will on equal footing. Don Ihde suggests that ANT is not genuinely symmetrical for the simple reason that the knowledge it represents is embodied in bodies, not things. It is not uncommon for ANT opponents to make phenomenological arguments about what they sense and feel that things are doing (or, rather, *not* doing), even when they would not usually subscribe to a phenomenological framework. They experience a basic difference between disagreeing with a chair and with a human being. I think we should take this kind of argument seriously. But instead of elevating it to a truth about the world—after all, some people do experience things as willing actors[99]—we should view it as an empirical phenomenon as we did with theories of self. It tells us something about how we typically imbue bodies with meaning and make them biopolitically productive. It tells us something about how power, respect, and recognition are produced. It is from this anthropocentric experience that care for persons is drawn. In turn, this care is the driving force for many of the biopolitical games that make cadavers useful and de-subjectify and exploit other living beings.

In short, bodies are poorly understood as static shrines of clearly delineated persons as some market opponents would have it; embedded in material flows, all of them will eventually fall back to the material world of which they form. Conversely, they are also poorly understood as material objects on par with stones, chairs, or coins as suggested by the market advocates who seek to separate subjectivity from materiality as person from body. Everything passing through the body acquires a potential connection to subjectivity. Through ubjects, we experiment with what persons are and how they should be treated. They present themselves for this experimentation thanks to partial connections to bodies of willful action. Different societies treasure different ubjects and exhibit care with a great variety of rituals. But irrespective of dominant cultural categorizations, every ubject possesses an unruly potential. For these reasons, ubjects host a fruitful darkness well worth exploring in more depth.

Endnotes

1. Lundin and Åkesson (1996).
2. Martin (1992).
3. Taylor (2005:749); see also Strathern and Lambek (1998) who insist on acknowledgement of phenomenological experience without reification of a transcendental body category. See also Latour's discussion of the ways in which knowing the body is constitutive for what bodies are and what can be done to them (Latour 2004).
4. The political project is often implicit; still many scholars seem to share Donna Haraway's view: "The perfection of the fully defended, 'victorious' self is a chilling fantasy" (Haraway 1991:224). I am influenced by elements of this political project, as I with the basic ontology, but I fear an ontology of emergence can have as many cruel effects as a stable and positivist one.
5. Ihde (2002) refers to the phenomenological body as Body One and the Foucauldian body as Body Two. For a discussion and critique, see Feenberg (2006). This resonates of course with the mindful body described by Nancy Scheper-Hughes and Margaret Lock (1987) as comprising a body-self (corresponding to Body One), a social body (explored by structuralist theory which I will return to in a discussion of categorization below), and a body politic (corresponding to Body Two).
6. A similar argument has been made by Brown et al. in their work on xenotransplantation. They write: "Crucially, mess is a consequence of purification and not a cause, a 'by-product' of ordering and for Latour, it is the very act of purification that proliferates the production of hybrids. Boundary-making is intended to deny connection, to foreclose the production of hybrids, and so paradoxically acts to facilitate their manufacture" (Brown et al. 2006a:209).
7. The proliferation of body imaginary that they facilitate, in conjunction with new medical technologies aimed at enhancement, has according to Andrew Webster made bodies seem more malleable than ever before (Webster 2006); see also Schicktanz (2007) and Shilling (1993).
8. Dickenson (2007), Holland (2001), and Kimbrell (1993). Lisa Blackman suggests that our major current challenge is to find ways to discuss integrity without presuming bodily boundaries (Blackman 2010), and while we might agree to some extent, the very posing of the challenge itself represents assumptions about integrity as tied to bounded entities and can thus be seen as an act of purification.
9. Hogle (1996).
10. Mol (2002) and Mol and Law (2004).

11. Mol and Law (2004:57).
12. Enfield et al. (2006).
13. Ihde criticizes Merleau-Ponty for working with an implicit sports body (Ihde 2002), and Feenberg (2006) in turn criticizes Ihde for not considering the dependent and the extended body. Partly, what I will do below is consider what that would imply in the sense that I discuss subjects (extended bodies on the move) and cadavers (dependents on the care of others to remain a body of meaning).
14. Merleau-Ponty (2002).
15. A parallel anthropological literature has followed Lock's invitation to avoid distinguishing between biology and culture. In her work on the different ways of undergoing menopause around the world, she illustrated the dangers of presuming nature to produce similar diseases which were simply "read" differently in divergent cultures (Lock 1993). We should instead approach bodies as states of being that are always culturally mediated (Scheper-Hughes and Lock 1987). See also Mauss (2007).
16. Farquhar and Lock (2007:2).
17. Farquhar and Lock (2007:15).
18. Merleau-Ponty (2002:431).
19. Epstein (1973). A similar but less explicit analytical shift is made when Wolputte reverts to the basic observation that it is common for "people [to] create or maintain a sense of self" Wolputte (2004:261). Merleau-Ponty (2002) would probably not like the notion of "theories" because of its connotations to pure cognition. I would suggest that a person's theory of self is enacted through concrete being-in-the-world and it is intuitive and embodied rather than abstract and contemplated, but the notion of theory serves the purpose of making it into a type of knowledge dependent on experience and practice, instead of than a pregiven capacity. The emphasis on cognition has been criticized also for ignoring insights from neurobiology (Quinn 2006), but while I acknowledge that there is no free-floating immaterial "sense of self," I wish to avoid subscribing to an ontologization of self as a neurobiological entity.
20. Shukin (2009).
21. Haraway (2008).
22. In some of the literature on personhood and subjecthood, it seems to be an implicit assumption that being recognized as a subject is *always* deemed desirable and deprivation of personhood *always* humiliating (see, e.g., Desjarlais 2000). However, we should recognize that objectification is not only a feature of many celebrated power structures such as the military where it is part of facilitating desired effects for the involved parties (irrespective of what I personally think of them). Objectification forms part of producing mechanisms associated with glory and honor. Objectification is also a desired aspect of many people's sexual life. Finally, it has been argued that reduction of bodies to "mere objects" rarely ever takes place—even if it is in some instances a sort of ideal which can never be reached (Cussins 1998; Latour 2004), though totalitarian regimes have indeed come dangerously close to total objectification of unwanted bodies.
23. A good introduction features in Carrithers et al. (1985). See also Bodenhorn and Bruck (2006) for a discussion of naming and naming practices as alternative entry points to relatedness and intersubjectivity; Wolputte (2004) for a general discussion of anthropological studies of the relationship between personhood, selfhood subjectivity, and the body; and Kaufman and Morgan (2005) for a discussion of typical contestations of what a person is.
24. Read (1955).
25. Bird-David (2004).
26. Leach (2005) and Weiss (1998).
27. Dreger (2004); see also Bratton and Chetwynd (2004) and Shildrick (1999). It is not uncommon for people to have an encapsulated twin in their body that they do not know about.
28. Martin (2007).
29. Kamper-Jørgensen et al. (2012).
30. Lock (2000, 2002) and Sharp (2000, 2007). Haddow has interviewed relatives after the donation and argues that the very act of donation stimulates thoughts about how persons relate to

bodies—basically, what a person is, if not the body that can be disintegrated, packaged, and implanted into others (Haddow 2005). See also Alnæs (2003), Fox and Swazey (1992), Hogle (1996), and Scheper-Hughes (2000).

31. Another example of alternative ways of relating to organs-ways that depart from the biomedical paradigm-came from a Swedish study finding that some elderly recipients like the idea of a "good match" to be facilitated with organs from people of their own age, though in medical terms they constitute "marginal donors" of lower quality (Idvall and Lundin 2007).

32. Sanal (2008). Some market proponents suggest that it could ease the feeling of being obliged toward the donor if you could *pay* for the organ (Satel 2008). This sort of obligation is related to what Fox and Swazey (1992) talk about as the tyranny of gift.

33. Hoeyer (2010).

34. Pfeffer (forthcoming) is currently undertaking important work aimed at uncovering many of these everyday transplant technologies that are rarely debated, focusing in particular on cornea and skin. Other transplant types stimulating limited public and ethical interest are dura mater (brain covering tissue) which has been transplanted since 1925 and arteries, tendons, and other tissue bits (Dexter 1965; Prolo 1981; Wilson 1947).

35. This same point could be made in relation to ubjects used in studies of assisted reproductive technologies (ART) (Franklin 1997; Strathern 1995; Thompson 2005); see also Clarke (1995, 1998) and Morgan (2002, 2003). For an anthropological study specifically emphasizing the agency of the embryo, see Konrad (1998).

36. See discussion in Chap. 1. I use the term in the following in ways closely related to what Strathern talks about as merographic connections in *After Nature,* but I continue to use the notion of partial connections to avoid overemphasizing the difference between the theorizing of anthropologists and their informants (Strathern 1995).

37. Strathern (2009:151).

38. Landecker (2007).

39. Brown (2009), Brown et al. (2006b, c), Eriksson and Webster (2008), Sperling (2004), Waldby (2006), Waldby and Squier (2004), and Williams et al. (2003). Susanne Lundin has interviewed research participants in xenotransplantation trials and found that from the patient's perspective it was not necessarily the technology that was seen as causing confusion; it could also be viewed as an attempt of creating order and wholeness in bodies experiencing disorder and disease (Lundin 2002).

40. The early announcement in the 1920s of an immortal chicken cell line was surrounded by quite a lot of attention, as an exception to the rule. It was much later found out that it was probably not immortal but constantly recreated with new cells (Skloot 2010). Duncan Wilson (2005) has argued that the early tissue culture research in the UK was actively engaging the media in attempts to raise expectations and get support for the research. This strategy failed, as the sociology of expectations has subsequently shown to be common (Brown and Michael 2003). Most cell lines remain unknown in the wider public.

41. Landecker (2007) and Skloot (2010). The complexity of the relationship between cell lines and their surroundings is also discussed in Landecker (1999) and Parry and Gere (2006).

42. Pfeffer and Laws (2006).

43. See, for example, Cambon-Thomsen (2004), Greely (2007), Kaye (2006), and Winickoff and Winickoff (2003). Important work by anthropologists, sociologists, and STS scholars has shown the differences between the policy framing of biobanks and the concerns of the donating public; see for example, Busby and Martin (2006), Ducournau (2007), Haddow et al. (2007), Pálsson and Rabinow (2005), Skolbekken et al. (2005), and Tutton (2007). I have written about such differences in Hoeyer (2006b) and provided a review of the biobank literature which criticizes its single-minded focus on informed consent in Hoeyer (2008).

44. Rabinow (1999) and Titmuss (1997). See Tutton (2004) for a discussion of the link between theorizing blood donation and biobanking.

45. Healy (2006). Healy faces a very different type of blood donation system, however, because today most blood is thoroughly processed and made into specialized products. In the process,

not only the blood but also the monetary aspects take on new forms. Before any blood products reach the veins of the recipients, many private actors have made a living on their manufacture and distribution. We are poorly equipped to understand this type of economy with the theoretical tools delivered in the pro/con market model literature, and Chap. 4 faces that challenge.

46. Copeman (2005, 2009).

47. Similarly, in a study from Denmark, Mette Nordahl Svendsen found that the actual couples donating embryos did not consider it fervently problematic to donate embryos for stem cell research. They found it more problematic to donate them to other infertile couples. For the couples, the relation constructed with the national welfare state through the donation was part of ingraining the research endeavors with legitimacy. By donating to research, the couples saw themselves contributing to a national community and to future public health efforts, which was in fact easier than becoming related to specific couples potentially giving birth to a child (Svendsen 2007). See also Simpson (2004).

48. Turkle (2008). Ott, for example, describes how artificial eyes have been applied differently to different classes: in the nineteenth century the poor were operated for preventive measures (so that a disease would not spread), while the "educated classes" could wait with operations until they were absolutely necessary (Ott 2002a, b). These decisions reflected a strong tradition for reading class relationships through bodily deficiencies; see also Blackwell (2004) for his work on class and body in the eighteenth century.

49. Serlin (2002); see also Kurzman (2003).

50. Faulkner and Kent (2001) and Kent and Faulkner (2002).

51. Frank (2006) and Haiken (2002).

52. Epidemiological studies of breast operations question the ability of bodily transformation to satisfy the need for self-confidence. Women who undergo breast reductions or enlargements suffer from increased mortality compared to the average population: the group undergoing enlargement due to many different diseases related also to so-called lifestyle choices but especially due to higher rates of suicide, while the group undergoing reductions performs better on all other health indicators relative to the average population but has increased mortality in relation to higher suicide rates (Jacobsen et al. 2004).

53. Kent (2003).

54. Haraway (2004). A lot of this debate has come to revolve around the relations people establish through computer-mediated technologies, and the cyborg figure has become a key entry point to discussions of virtual reality (Ihde 2002). Here, however, I focus on traveling ubjects.

55. On using biology to probe our thinking, see also the discussion in Grosz (2004).

56. http://nihroadmap.nih.gov/hmp/. See Turnbaugh et al. (2007) for a description of the project and McGuire et al. (2008) and Nerlich and Hellsten (2009) for analyses of its social and ethical implications.

57. Mai and Draganov (2009) and Turnbaugh et al. (2009).

58. Margulis and Sagan (1987).

59. Hird (2009:84).

60. Haraway (1991:207).

61. Kauffman (1970:258–9).

62. Rabinow (1993). Rabinow's oft-cited article on Galton's regret, in which he made the argument that Galton was disappointed about the inability of fingerprinting to deliver a stable proof from which typologies of humans could be deducted, has later been criticized for ignoring the fact that attempts of linking fingerprints to social categories such as race and homosexuality continued with some amount of "success" well into the 1990s (Cole 2004). On forensic uses of genetics, see Cho and Sankar (2004), Derksen (2000), Lazer (2004), and Williams and Johnson (2004), and for a historical note on anthropology's own early attempts of defining the essence of humanity, see Hecht (2003).

63. Keller (2000). The issue of informatization has been explored by Waldby, Parry, and Tacker, among others (Parry 2004a, b; Thacker 2005; Waldby 2000), and the historical junctures involved are discussed in Garlick (2006). One of the reasons for me to focus on the ubject is to counterbalance the interest in informatization and focus on the materiality of the ubjects.

64. The results of the public and private effort were published in the International Human Genome Sequencing Consortium (2001) and Venter et al. (2001). For a discussion of the subsequent challenges to the finitude of the findings, see, for example, Lock (2005), Marks (2003), and Noble (2006). The implications of the tendency for genetic determinism have been discussed at great length in the social sciences; see, for example, Finkler et al. (2003), Hedgecoe (2000), and Rothstein (2005).

65. Dear (2009).

66. Kristeva (1982:3).

67. Green (2008:72).

68. Merleau-Ponty (2002:140).

69. She writes: "Therefore the categories of subject and object are not antipoles; rather, paraphrasing Latour, they come together as hybrids, which is a fruitful way of looking at the body parts of organ donors" (Jensen 2010:77).

70. As discussed in Cooter (2000); see also important contributions on the political lives of dead bodies in Timmermans (2006) and Verdery (1999). The classical collection of death rites is to be found in Hertz (1960).

71. Frazer (1993 [1890]:233).

72. For cannibalism, see Poole (1983) and Sahlins (1983), and for religious fetish, see Ellis (2002).

73. Leach (1958:157) and Douglas (1995:chapter 7). In an interesting comment, Sjaak van der Geest (2007) suggests that these classic works, as well as anthropology in general, have evaded exploration of the handling of one of the most basic of ubjects: feces.

74. For a discussion of using substances as means of communication and relation building, see Hutchinson (2000).

75. Durkheim (2008:137).

76. Douglas (1995:12).

77. Turner (1967), quotes from pages 102, 96–97, and 110.

78. Squier (2004).

79. Enfield et al. (2006), Hillman and Mazzio (1997), and Norri (1998).

80. Elias (1994). Mauss has also written on the topic of embodiment of social structures (Mauss 2007) using the notion of *habitus* long before Bourdieu (2000) made it common parlance. Another contribution to the understanding of embodiment of historical shifts is Connerton's work on bodily memory (Connerton 1989).

81. Laporte (2000). Along similar lines, Allen points out how farting provokes by way of bringing attention to the lack of clear body boundaries (Allen 2010).

82. Cromley (1990) and Cromley (1996). If one were to write this history of changed notions of bodily purification, it would be important to include not only the historical rise of measures of hygiene (Armstrong 2002; Bashford 2004) but also the emergence of commodities such as dispensable sanitary towels and other intimate products facilitating self-purification while creating commercial value (Shail 2007). There is perhaps yet an interesting dimension to the story of coproduction and mutual interdependence between changes in management of body "waste" and the rise of the capitalist mode of production: Abelove suggests that the population rise in the UK in 1680–1830 that produced a crucial labor reserve for industrialization can only be explained with changed understandings of sexual relations. All acts that did not aim at the direct deposit of male sperm in a female vagina became abject, and as a consequence, more babies were born (Abelove 2007).

83. Schlick (1966).

84. Morris (1991).

85. Schopenhauer (2004, 2006). This notion of will resonates with mainstream social science in as far as it is dealing with motivation and agency in manners that do not reduce motivation to intention and action to logocentric planning (Hastrup 1995). It makes an odd fit with most moral philosophy, however, where the focus tends to be on "*free*" will" as a cognitive capacity to choose between available options and bear the responsibility for one's choice (Hoeyer and Lynöe 2006). In my view, responsibility need not presuppose intentional, conscious choice, but

my point here is not about responsibility as such, but about the basic experience of will and its importance for the phenomenological experience of subjecthood.
86. Nietzsche (2000:219).
87. Nietzsche (1967:481). Note how well this compares to Merleau-Ponty's attack on Kant and Descartes: "Truth does not 'inhabit' only 'the inner man', or more accurately, there is no inner man, man is in the world, and only in the world does he know himself" (Merleau-Ponty 2002:xii).
88. Nietzsche (2000:218).
89. Jackson (1998, 2002); see also his general introduction to phenomenological anthropology in Jackson (1996).
90. Jackson (1998:24).
91. Sometimes people feel very strongly that something is not-me, though the medical profession does not acknowledge their longings. In Carl Elliott's (2003) work, we learn about the amputees by choice movement lobbying to have limbs amputated for the simple reason that they do not feel that a particular arm or leg is part of them. Such cases only underline the impossibility of establishing a universal norm for bodies but also how body norms are established in a wider biopolitical context.
92. Canguilhem (1978). Nikolas Rose has made a very different reading of Canguilhem in his book *The Politics of Life Itself*, but I find Canguilhem much more cautious and balanced than Rose seems to imply (Rose 2007).
93. Canguilhem (1978:132). This basic insight is often sidestepped, however, in search of a secure base in "nature" from which we might derive value judgments. One of the most famous attempts of a natural concept of disease is stated in Boorse (1977) used by, among others, Norman Daniels (1982, 1988) in his laudable attempts of arguing for a right to healthcare.
94. Canguilhem (1978:82).
95. Canguilhem (1978:175).
96. Elliott (2003).
97. Mol and Law (2004:45).
98. Wittgenstein (2001).
99. Turkle (1984). Also, mortuary rituals for objects are not uncommon and are sometimes related to fear of objects returning and doing harm (Kretschmer 2000). Objects can indeed be phenomenologically experienced as having a will.

References

Abelove H (2007) some speculations on the history of "sexual intercourse" during the "long eighteenth century" in England. In: Lock M, Farquhar J (eds) Beyond the body proper: reading the anthropology of maternal life. Duke University Press, Durham/London, pp 217–223

Allen V (2010) On farting. Language and laughter in the middle ages. Palgrave Macmillan, New York

Alnæs AH (2003) The anthropological fieldworker in a biomedical high-tech setting: some methodological problems and experiences. J Appl Anthropol Policy Pract 10(3):9–18

Armstrong D (2002) A new history of identity: a sociology of medical knowledge. Palgrave, London

Bashford A (2004) Imperial hygiene: a critical history of colonialism, nationalism and public health. Palgrave Macmillan, New York

Bird-David N (2004) Illness-images and joined beings: a critical/Nayaka perspective on intercorporeality. Soc Anthropol 12(3):325–339

Blackman L (2010) Bodily integrity. Body Soc 16(3):1–9

Blackwell M (2004) "Extraneous bodies": the contagion of live-tooth transplantation in late-eighteenth-century England. Eighteenth-Century Life 28(1):21–68

Bodenhorn B, Bruck GV (2006) "Entangled in histories": an introduction to the anthropology of names and naming. In: Bruck GV, Bodenhorn B (eds) The anthropology of names and naming. Cambridge University Press, Cambridge, pp 1–30

Boorse C (1977) Health as a theoretical concept. Philos Sci 44(4):542–573

Bourdieu P (2000) Outline of a theory of practice. Cambridge University Press, Cambridge

Bratton MQ, Chetwynd SB (2004) One into two will not go: conceptualising conjoined twins. J Med Ethics 30:279–285

Brown N (2009) Beasting the embryo: the metrics of humanness in the transpecies embryo debate. BioSocieties 4:147–163

Brown N, Michael M (2003) A sociology of expectations: retrospecting prospects and prospecting retrospects. Technol Anal Strateg Manage 15(1):3–18

Brown N, Faulkner A, Kent J, Michael M (2006a) Regulating hybridity: policing pollution in tissue engineering and transpecies transplantation. In: Webster A (ed) New technologies in health care: challange, change and innovation. Palgrave Macmillan, Hampshire, pp 194–210

Brown N, Faulkner A, Kent J, Michael M (2006b) Regulating hybrids: "making a mess" and "cleaning up" in tissue engineering and transpecies transplantation. Soc Theory Health 4:1–24

Brown N, Kraft A, Martin P (2006c) The promissory pasts of blood stem cells. BioSocieties 1:329–348

Busby H, Martin P (2006) Biobanks, national identity and imagined communities: the case of UK Biobank. Sci Cult 15(3):237–251

Cambon-Thomsen A (2004) The social and ethical issues of post-genomic human biobanks. Nat Rev Genet 5:6–13

Canguilhem G (1978) On the normal and the pathological. D. Reidel Publishing Company, Dordrecht

Carrithers M, Collins A, Lukes S (1985) The category of the person: anthropology, philosophy, history. Cambridge University Press, Cambridge

Cho MK, Sankar P (2004) Forensic genetics and ethical, legal and social implications beyond the clinic. Nat Genet Suppl 36(11):8–12

Clarke AE (1995) Research materials and reproductive science in the United States, 1910–1940. In: Star SL (ed) Ecologies of knowledge. State University of New York, Albany, pp 183–225

Clarke AE (1998) Disciplining reproduction, modernity, American life sciences, and "the problems of sex". University of California Press, London

Cole SA (2004) Fingerprint identification and the criminal justice system: historical lessons for the DNA debate. In: Lazer D (ed) DNA and the criminal justice system: the technology of justice. The MIT Press, Cambridge, MA, pp 63–89

Connerton P (1989) How societies remember. Cambridge University Press, Cambridge

Cooter R (2000) The dead body. In: Cooter R, Pickstone J (eds) Medicine in the twentieth century. Harwood Academic Publishers, Amsterdam, pp 469–485

Copeman J (2005) Veinglory: exploring processes of blood transfer between persons. J R Anthropol Inst 11:465–485

Copeman J (2009) Veins of devotion: blood donation and religious experience in North India. Rutgers University Press, New Brunswick

Cromley E (1990) Sleeping around: a history of American beds and bedrooms. The second Banham memorial lecture. J Des Hist 3(1):1–17

Cromley E (1996) Transforming the food axis: houses, tools, modes of analysis. Mater Hist Rev 44(Fall):8–22

Cussins CM (1998) Ontological choreography: agency for women patients in an infertility clinic. In: Berg M, Mol A (eds) Differences in medicine: unraveling practices, techniques, and bodies. Duke University Press, Durham/London, pp 166–201

Daniels N (1982) Health-care need and distributive justice. In: Cohen M, Nagel T, Scanlon T (eds) Medicine and moral philosophy. Princeton University Press, Princeton, pp 81–114

Daniels N (1988) Justice in health care. Am i my parents' keeper?—an essay on justice between the young and the old. Oxford University Press, Oxford, pp 66–82

Dear PH (2009) Copy-number variation: the end of the human genome? Trends Biotechnol 27(8):448–454

Derksen L (2000) Towards a sociology of measurement: the meaning of measurement error in the case of DNA profiling. Soc Stud Sci 30(6):803–845

Desjarlais R (2000) The makings of personhood in a shelter for people considered homeless and mentally ill. Ethos 27(4):466–489

Dexter F (1965) The preservation of tissue for surgical transplantation and subsequent formation of a tissue bank. J Sci Technol 11(4):149–176

Dickenson D (2007) Property in the body: feminist perspectives. Cambridge University Press, New York

Douglas M (1995) Purity and danger: an analysis of the concepts of pollution and taboo. Routledge, London

Dreger AD (2004) One of us: conjoined twins and the future of normal. The President and Fellows of Harvard College, Harvard

Ducournau P (2007) The viewpoint of DNA donors on the consent procedure. New Genet Soc 26(1):105–116

Durkheim E (2008) The elementary forms of the religious life. Dover, Mineola

Elias N (1994) The history of manners: the civilizing process. Blackwell, Oxford, p vii-255

Elliott C (2003) Better than well: American medicine meets the American dream. Norton & Company, New York

Ellis B (2002) Why is a lucky rabbit's foot lucky? Body parts as fetishes. J Folk Res 39(1):51–84

Enfield NJ, Majid A, Staden MV (2006) Cross-linguistic categorisation of the body: introduction. Lang Sci 28:137–147

Epstein S (1973) The self-concept revisited, or a theory of a theory. Am Psychol 28:404–416

Eriksson L, Webster A (2008) Standardizing the unknown: practicable pluripotency as doable futures. Sci Cult 17(1):57–69

Farquhar J, Lock M (2007) Introduction. In: Lock M, Farquhar J (eds) Beyond the body proper: reading the anthropology of material life. Duke University Press, Durham, pp 1–16

Faulkner A, Kent J (2001) Innovation and regulation in human implant technologies: developing comparative approaches. Soc Sci Med 53:895–913

Feenberg A (2006) Active and passive bodies: Don Ihde's phenomenology of the body. In: Selinger E (ed) Postphenomenology: a critical companion to Ihde. State University of New York Press, Albany, pp 189–196

Finkler K, Skrzynia C, Evans JP (2003) The new genetics and its consequences for family, kinship, medicine and medical genetics. Soc Sci Med 57:403–412

Fox RC, Swazey JP (1992) Spare parts: organ replacement in American Society. Oxford University Press, New York

Frank K (2006) Agency. Anthropol Theory 6(3):281–302

Franklin S (1997) Embodied progress: a cultural account of assisted conception. Routledge, London

Frazer J (1993) The golden bough. Wordsworth Editions Ltd, Cumberland House, Hertfordshire

Garlick S (2006) Mendle's generation: molecular sex and the informatic body. Body Soc 12(4):53–71

Greely HT (2007) The uneasy ethical and legal underpinnings of large-scale genomic biobanks. Annu Rev Genomics Hum Genet 8:343–364

Green JW (2008) Beyond the good death: the anthropology of modern dying. University of Pennsylvania Press, Philadelphia

Grosz E (2004) The nick of time: politics, evolution and the untimely. Duke University Press, Durham/London

Haddow G (2005) The phenomenology of death, embodiment and organ transplantation. Sociol Health Illn 27(1):92–113

Haddow G, Laurie G, Cunningham-Burley S, Hunter KG (2007) Tackling community concerns about commercialisation and genetic research: a modest interdisciplinary proposal. Soc Sci Med 64:272–282

Haiken E (2002) Modern miracles: the development of cosmetic prosthetics. In: Ott K, Serlin D, Mihm S (eds) Artificial parts, practical lives: modern histories of prosthetics. New York University Press, New York, pp 171–198

Haraway DJ (1991) The biopolitics of postmodern bodies: constitutions of self in immune systems discourse. In: Simians, cyborgs, and women: the reinvention of nature. Free Association Books, London, pp 203–254

Haraway DJ (2004) A manifesto for cyborgs: science, technology, and socialist feminism in the 1980s. In: The Haraway reader. Routledge, London, pp 7–45

Haraway DJ (2008) When species meet. University of Minnesota Press, Minneapolis

Hastrup K (1995) A passage to anthropology: between experience and theory. Routledge, London

Healy K (2006) Last best gift: altruism and the market for human blood and organs. The University of Chicago Press, Chicago

Hecht JM (2003) The end of the soul: scientific modernity, atheism, and anthropology in France. Columbia University Press, New York

Hedgecoe AM (2000) Essay review: the popularization of genetics as geneticization. Public Underst Sci 9:183–189

Hertz R (1960) Death and the right hand. Cohen and West, London

Hillman D, Mazzio C (1997) Introduction: individual parts. In: Hillman D, Mazzio C (eds) The body in parts: fantasies of corporeality in early modern Europe. Routledge, London, pp xi–xxix

Hird MJ (2009) The origins of sociable life: evolution after science studies. Palgrave Macmillan, Hampshire

Hoeyer K (2006) The power of ethics: a case study from Sweden on the social life of moral concerns in policy processes. Sociol Health Illn 28(6):785–801

Hoeyer K (2008) The ethics of research biobanking: a critical review of the literature. Biotechnol Genet Eng Rev 25:429–452

Hoeyer K (2010) After novelty: the mundane practices of ensuring a safe and stable supply of bone. Sci Cult 19(2):123–150

Hoeyer K, Lynöe N (2006) Motivating donors to genetic research? Anthropological reasons to rethink the role of informed consent. Med Health Care Philos 9:13–23

Hogle L (1996) Transforming "body parts" into therapeutic tools: a report from Germany. Med Anthropol Q 10(4):675–682

Holland S (2001) Beyond the embryo: a feminist appraisal of the embryonic stem cell debate. In: Holland S, Lebacqz K, Zoloth L (eds) The human embryonic stem cell debate. science, ethics, and public policy. The MIT Press, Cambridge/London, pp 73–86

Hutchinson SE (2000) Identity and substance: the broadening bases of relatedness among the Nuer of southern Sudan. In: Carsten J (ed) Cultures of relatedness: new approaches to the study of kinship. Cambridge University Press, Cambridge, pp 55–72

Idvall M, Lundin S (2007) Transplantation with kidneys from marginal donors. Ethnol Scand 37:1–18

Ihde D (2002) Bodies in technology. University of Minnesota Press, Minneapolis

International Human Genome Sequencing Consortium (2001) Initial sequencing and analysis of the human genome. Nature 409:860–921

Jackson M (1996) Introduction: phenomenology, radical empiricism, and anthropological critique. In: Jackson M (ed) Things as they are: new directions in phenomenological anthropology. Indiana University Press, Bloomington, pp 1–50

Jackson M (1998) Minima ethnographica: intersubjectivity and the anthropological project. University of Chicago Press, Chicago

Jackson M (2002) Familiar and foreign bodies: a phenomenological exploration of the human-technology interface. J R Anthropol Soc 8:333–346

Jacobsen PH, Hölmich LR, McLaughlin JK, Johansen C, Olsen JH, Kjøller K, Friis S (2004) Mortality and suicide among Danish women with cosmetic breast implants. Arch Intern Med 164:2450–2455

Jensen AMB (2010) A sense of absence: the staging of heroic deaths and ongoing lives among American organ donor families. In: Bille M, Hastrup F, Sørensen TF (eds) The anthropology of absence: materializations of transcendence and loss. Springer, New York, pp 68–84

Kamper-Jørgensen M, Biggar RJ, Tjønneland A, Hjalgrim H, Kroman N, Rostgaard K, Stamper CL, Olsen A, Andersen AM, Gadi VK (2012) Opposite effects of microchimerism on breast and colon cancer. Eur J Cancer 48(14):2227–2235

Kauffman SA (1970) Articulation of parts explanation in biology and the rational search for them. Proc Bienn Meet Soc Sci Assoc 1970:257–272

Kaufman SR, Morgan LM (2005) The anthropology of the beginnings and ends of life. Annu Rev Anthropol 34:317–341

Kaye J (2006) Police collection and access to DNA samples. Genomics Soc Policy 2(1):16–27

Keller EF (2000) The century of the gene. Harvard University Press, Cambridge, MA

Kent J (2003) Lay experts and the politics of breast implants. Public Underst Sci 12(4):403–421

Kent J, Faulkner A (2002) Regulating human implant technologies in Europe —understanding the New Era in medical device regulation. Health Risk Soc 4(2):189–209

Kimbrell A (1993) The human body shop: the engineering and marketing of life. Harper Collins Religious, London

Konrad M (1998) Ova donation and symbols of substance: some variations on the theme of sex, gender and the partible body. J R Anthropol Inst 12(1):643–667

Kretschmer A (2000) Mortuary rites for inanimate objects: the case of Hari Kuyō. Jap J Relig Stud 27(3–4):379–404

Kristeva J (1982) Powers of horror: an essay on abjection. Columbia University Press, New York

Kurzman S (2003) Performing able-bodiedness: amputees and prosthetics in America. University of California, Santa Cruz

Landecker H (1999) Between beneficence and chattel: the human biological in law and science. Sci Context 12(1):203–225

Landecker H (2007) Culturing life: how cells became technologies. Harvard University Press, Cambridge

Laporte D (2000) History of shit. The MIT Press, Cambridge, MA

Latour B (2004) How to talk about the body? The normative dimension of science studies. Body Soc 10(2–3):205–229

Lazer D (2004) Introduction: DNA and the criminal justice system. In: Lazer D (ed) DNA and the criminal justice system: the technology of justice. The MIT Press, Cambridge, MA, pp 3–12

Leach ER (1958) Magical hair. J R Anthropol Inst G B Irel 88(2):147–164

Leach J (2005) Livers and lives: organ extraction narratives on the Rai coast of Papua New Guinea. In: van Binsbergen WMJ, Geschiere PL (eds) Commodification: things, agency, and identities. Lit Verlag, Münster, pp 283–300

Lock M (1993) Encounters with aging: mythologies of menopause in Japan and North America. University of California Press, Berkeley

Lock M (2000) On dying twice: culture, technology and the determination of death. In: Lock M, Young A, Cambrosio A (eds) Living and working with the new medical technologies: intersections of inquiry. Cambridge University Press, Cambridge, pp 233–262

Lock M (2002) Twice dead: organ transplants and the reinvention of death. University of California Press, Berkley

Lock M (2005) Eclipse of the gene and the return of divination. Curr Anthropol 46(suppl):47–70

Lundin S (2002) Creating identity with biotechnology: the xenotransplanted body as norm. Public Underst Sci 11:333–345

Lundin S, Åkesson L (1996) Introduction. In: Lundin S, Åkesson L (eds) Bodytime: on the interaction of body, identity, and society. Lund University Press, Lund, pp 5–12

Mai V, Draganov PV (2009) Recent advances and remaining gaps in our knowledge of associations between gut microbiota and human health. World J Gastroenterol 15(1):81–85

Margulis L, Sagan D (1987) Microcosmos: four billion years of evolution from our microbial ancestors. Allen & Unwin, London

Marks J (2003) 98% chimpanzee and 35% daffodil: the human genome in evolutionary and cultural context. In: Goodman A, Heath D, Lindee S (eds) Genetic nature/culture: anthropology and science beyond the two-culture divide. University of California Press, Berkeley, pp 132–152

Martin E (1992) The end of the body? Am Ethnol 19(1):121–140

Martin A (2007) The Chimera of liberal individualism: how cells became selves in human clinical genetics. In: Eghigian G, Killen A, Leuenberger C (eds) The self as project—politics and the human sciences. University of Chicago Press, Chicago, pp 206–222

Mauss M (2007) Techniques of the body. In: Lock M, Farquhar J (eds) Beyond the body proper: reading the anthropology of maternal life. Duke University Press, Durham/London, pp 50–68

McGuire AL, Colgrove J, Whitney SN et al (2008) Ethical, legal, and social considerations in conducting the Human Microbiome Project. Genome Res 18:1861–1864

Merleau-Ponty M (2002) Phenomenology of perception. Routledge, London

Mol A (2002) The body multiple: ontology in medical practice. Duke University Press, London

Mol A, Law J (2004) Embodied action, enacted bodies: the example of hypoglycaemia. Body Soc 10(2/3):43–62

Morgan L (2002) "Properly disposed of": a history of embroyo disposal and the changing claims on fetal remains. Med Anthropol 21(3):247–274

Morgan L (2003) Embroyo tales. In: Franklin S, Lock M (eds) Remaking life and death: toward and anthropology of the biosciences. School of American Research Press/James Currey, Santa Fe, pp 261–291

Morris B (1991) Western Conceptions of the Individual. Berg, Oxford

Nerlich B, Hellsten I (2009) Beyond the human genome: microbes, metaphors and what it means to be human in an interconnected post-genomic world. New Genet Soc 28(1):19–36

Nietzsche F (1967) On the genealogy of morals. The Modern Library, New York

Nietzsche F (2000) Beyond good and evil: prelude to a philosophy of the future. In: Kaufmann W (ed) Basic writings of Nietzsche. Random House, New York, pp 179–435

Noble D (2006) The music of life: biology beyond the genome. Oxford University Press, Oxford

Norri J (1998) Names of body parts in English, 1400–1550. Anneles Academiae Scientiarum Fenicae, Helsinki

Ott K (2002a) Hard wear and soft tissue: craft and commerce in artificial eyes. In: Ott K, Serlin D, Mihm S (eds) Artificial parts, practical lives: modern histories of prosthetics. New York University Press, New York, pp 147–170

Ott K (2002b) The sum of its parts: an introduction to modern histories of prosthetics. In: Ott K, Serlin D, Mihm S (eds) Artificial parts, practical lives: modern histories of prosthetics. New York University Press, New York, pp 1–42

Pálsson G, Rabinow P (2005) The Iceland controversy: reflections on the trans-national market of civic virtue. In: Ong A, Collier S (eds) Global assemblages: technology, politics, and ethics as anthropological problems. Blackwell, Oxford, pp 91–103

Parry B (2004a) Bodily transactions: regulating a new space of flows in "bio-information". In: Verdery K, Humphrey C (eds) Property in question: value transformation in the global economy. Berg, Oxford, pp 29–68

Parry B (2004b) Trading the genome: investigating the commodification of bio-information. Columbia University Press, New York/Chichester/West Sussex

Parry B, Gere C (2006) Contested bodies: property models and the commodification of human biological artefacts. Sci Cult 15(2):139–158

Pfeffer N (forthcoming) Insider trading. Yale University Press, London

Pfeffer N, Laws S (2006) "It's only a blood test": what people know and think about venepuncture and blood. Soc Sci Med 62(12):3011–3023

Poole FJP (1983) Cannibals, tricksters, and witches: anthropophagic images among Bimin-Kuskusmin. In: Brown P, Tuzin D (eds) The ethnography of cannibalism. Society for Psychological Anthropology, Washington, DC, pp 6–32

Prolo DJ (1981) Use of transplantable tissue in neurosurgery. In: Carmel PW (ed) Clinical neuro-surgery: proceedings of the congress of neurological surgeons, Houston, Texas 1980. Williams & Wilkins, Baltimore, pp 407–417

Quinn N (2006) The self. Anthropol Theory 6(3):362–384

Rabinow P (1993) Galton's regret and DNA typing. Cult Med Psychiatry 17(1):59–65

Rabinow P (1999) French DNA: trouble in purgatory. University of Chicago Press, Chicago

Read KE (1955) Morality and the concept of the person among the Gahuku-Gama. Oceania 25(4):233–282

Rose N (2007) The politics of life itself: biomedicine, power, and subjectivity in the twenty-first century. Princeton University Press, Princeton

Rothstein MA (2005) Genetic exceptionalism and legislative pragmatism. Hastings Cent Rep 35(4):27–33

Sahlins M (1983) Raw women, cooked men, and other "great things" of the Fiji Islands. In: Brown P, Tuzin D (eds) The ethnography of cannibalism. Society for Psychological Anthropology, Washington, DC, pp 72–93

Sanal A (2008) The dialysis machine. In: Turkle S (ed) The inner history of devices. The MIT Press, Cambridge, MA, pp 138–152

Satel S (2008) Concerns about human dignity and commodification. In: Satel S (ed) When altruism isn't enough: the case for compensating kidney donors. The AEI Press, Washington, DC, pp 63–78

Scheper-Hughes N (2000) The global traffic in human organs. Curr Anthropol 41(2):191–224

Scheper-Hughes N, Lock M (1987) The mindful body: a prolegomenon to future work in medical anthropology. Med Anthropol Q 1(1):6–41

Schicktanz S (2007) Why the way we consider the body matters— reflections on four bioethical perspectives on the human body. Philos Ethics Humanit Med 2:30–41

Schlick M (1966) When is a man responsible? In: Berofsky B (ed) Free will and determinism. Harper & Row, London, pp 54–63

Schopenhauer A (2004) On the suffering of the world. Penguin Books Ltd., London

Schopenhauer A (2006) Verden som vilje og forestilling. Gyldendals Bogklubber, København

Serlin D (2002) Engineering masculinity: veterans and prosthetics after World War Two. In: Ott K, Serlin D, Mihm S (eds) Artificial parts, practical lives: modern histories of prosthetics. New York University Press, New York, pp 45–74

Shail A (2007) "Although a woman's article": menstruant economics and creative waste. Body Soc 13(4):77–96

Sharp LA (2000) The commodification of the body and its parts. Annu Rev Anthropol 29:287–328

Sharp LA (2007) Bodies, commodities, and biotechnologies: death, mourning, and scientific desire in the realm of human organ transfer. Columbia University Press, New York

Shildrick M (1999) This body which is not one: dealing with differencess. Body Soc 5(2–3):77–92

Shilling C (1993) The body and social theory. Sage Publications, London

Shukin N (2009) Animal capital: rendering life in biopolitical times. University of Minnesota Press, Minneapolis

Simpson B (2004) Impossible gifts: bodies, Buddhism and bioethics in contemporary Sri Lanka. R Anthropol Inst 10(4):839–859

Skloot R (2010) The immortal life of Henrietta Lacks. Crown Publishing Group, New York

Skolbekken J-A, Ursin LØ, Solberg B, Christensen E, Ytterhus B (2005) Not worth the paper it's written on? Informed consent and biobank research in a Norwegian context. Crit Public Health 15(4):335–347

Sperling S (2004) From crisis to potentiality. Sci Public Policy 31(2):139–149

Squier SM (2004) Liminal lives. Duke University Press, Durham

Strathern M (1995) After nature: English kinship in the late Twentieth Century. Cambridge University Press, Cambridge

Strathern M (2009) Using bodies to communicate. In: Lambert H, McDonald M (eds) Social bodies. Berghahn Books, New York, pp 148–169

Strathern A, Lambek M (1998) Introduction—embodying sociality: Africanist-Melanesianist comparisons. In: Lambek M, Strathern A (eds) Bodies and persons: comparative perspectives from Africa and Melanesia. Cambridge University Press, Cambridge, pp 1–25

Svendsen MN (2007) Between reproductive and regenerative medicine: practising embryo donation and civil responsibility in Denmark. Body Soc 13(4):21–45

Taylor JS (2005) Surfacing the body interior. Annu Rev Anthropol 34:741–756

Thacker E (2005) The global genome: biotechnology, politics, and culture. The MIT Press, Cambridge, MA

Thompson C (2005) Making parents: the ontological choreography of reproductive technologies. The MIT Press, Cambridge, MA

Timmermans S (2006) Postmortem: how medical examiners explain suspicious deaths. Chicago University Press, Chicago

Titmuss R (1997) The gift relationship: from human blood to social policy. The New Press, New York

Turkle S (1984) Thinking of yourself as a machine: the second self: computers and the human spirit. Granada Publishing, London, pp 281–318

Turkle S (2008) Inner history. In: Turkle S (ed) The inner history of devices. The MIT Press, Cambridge, MA, pp 2–29

Turnbaugh PJ, Ley RE, Hamady M, Fraser-Liggett CM, Knight R, Gordon JI (2007) The human microbiome project. Nature 449:804–810

Turnbaugh PJ, Hamady M, Yatsunenko T, Cantarel BL, Duncan A, Ley RE, Sogin ML, Jones WJ, Roe BA, Affourtit JP, Egholm M, Henrissat B, Heath AC, Knight R, Gordon JI (2009) A core gut microbiome in obese and lean twins. Nature 457(7228):480–484

Turner V (1967) Betwixt and between: the liminal period in rites de passage. In: The forest of symbols: aspects of Ndembu Ritual. Cornell University Press, Ithaca, pp 93–111

Tutton R (2004) Person, property and gift: exploring languages of tissue donation to biomedical research. In: Tutton R, Corrigan O (eds) Genetic databases: socio-ethical issues in the collection and use of DNA. Routledge, London

Tutton R (2007) Constructing participation in genetic databases: citizenship, governance and ambivalence. Sci Technol Hum Values 32(2):172–195

Van der Geest S (2007) Not knowing about defecation. In: Littlewood R (ed) On knowing and not knowing in medical anthropology. Left Coast Press, Walnut Creek, pp 75–86

Venter JC et al (2001) The sequence of the human genome. Science 291:1304–1351

Verdery K (1999) Dead bodies animate the study of politics. In: The political lives of dead bodies— reburial and postsocialist change. Columbia University Press, New York, pp 23–53

Waldby C (2000) The visible human project: informatic bodies and posthuman medicine. Routledge, London

Waldby C (2006) Umilical cord blood: from social gift to venture capital. BioSocieties 1:55–70

Waldby C, Squier SM (2004) Ontogeny, ontology, and phylogeny: embryonic life and stem cell technologies. Configurations 11:27–46

Webster A (2006) Introduction: new technologies in health care: opening the black bag. In: Webster A (ed) New technologies in health care: challange, change and innovation. Palgrave Macmillan, Hampshire, pp 1–8

Weiss B (1998) Electric vampires: Haya rumors of the commodified body. In: Lambek M, Strathern A (eds) Bodies and persons: comparative perspectives from Africa and Melanesia. Cambridge University Press, Cambridge, pp 172–194

Williams R, Johnson P (2004) Circuits of surveillance. Surveill Soc 2(1):1–14

Williams C, Kitzinger J, Henderson L (2003) Envisaging the embryo in stem cell research: rhetorical strategies and media reporting on the ethical debates. Sociol Health Illn 25(7):793–814

Wilson PD (1947) Experiences with a Bone Bank. Ann Surg 126(6):932–946

Wilson D (2005) The early history of tissue culture in Britain: the interwar years. Soc Hist Med 18(2):225–243

Winickoff DE, Winickoff RN (2003) The charitable trust as a model for genomic biobanks. N Engl J Med 349(12):1180–1184

Wittgenstein L (2001) Philosophical investigations: the German text with a revised English translation. Blackwell, London

Wolputte SV (2004) Hang on to your self: of bodies, embodiment, and selves. Annu Rev Anthropol 33:251–269

Chapter 4
Ubject Exchange as Everyday Practice

The two previous chapters provided reasons to reconsider what we mean when we casually talk about "markets" and "body parts." This chapter changes the scene by moving on to concrete studies of the everyday practices of ubject exchange. As we turn to the level of mundane everyday practices, we also move into a wetland of ambiguity in a quite different sense. It is in the concrete exchange practices that an ubject's unruly potential for subjecthood is doing its work. I show how ubject exchange can be analyzed without the dominant presumptions of "market forces" characterizing market thinking and without settling the ontological status of the ubjects in question. Also, I have a more specific agenda. I wish to highlight how moral agency is of key importance for the workings of ubject exchange.

I present work from two empirical studies of my own, one conducted in Sweden and one in Denmark. The Swedish study focuses on blood used for genetic research, while the Danish one focuses on hips used for implants. The blood is primarily desired for the information it contains, while the hips are primarily desired for their material-therapeutic properties, and I will show how such differences influence the valuation of the respective ubjects. The two types of ubject instigate different types of moral reasoning, but there are also some similarities across both settings in the form of moral concerns stimulated by the ubjects. Such concerns have important implications for what can acquire a "price" and how a "price" is acquired. The ambiguous status of the ubject does not prevent exchanges from taking place, but it makes these exchanges take on forms which are supposed to distinguish them from what the involved people associate with a "market." Price setting is a central feature of market thinking, as described in Chap. 2, but I will show how the mechanisms at play in ubject exchange can be very different from what market thinking would prescribe.[1] Once we move beyond market thinking, whether we find it morally appealing or appalling, it is easier to appreciate what people actually do when they exchange ubjects—without projecting particular forms of calculative agency onto their acts. And once we are less focused on how ubjects *ought* to be exchanged, either as things or body parts, it is also easier to see what the involved actors find gratifying or problematic about their exchange.

K. Hoeyer, *Exchanging Human Bodily Material: Rethinking Bodies and Markets*,
DOI 10.1007/978-94-007-5264-1_4, © Springer Science+Business Media Dordrecht 2013

Since I wish to take moral agency seriously, and since ethics policies are often mentioned as the prime way to construct proper ubject exchange systems, a few remarks on ethics as policymaking is needed. The literature on ubject exchange often takes one of two positions in relation to ethics. On the one hand, some (often ethicists) turn to ethics policies as a safeguard against the self-interest that market thinking is seen to bolster.[2] On the other hand, some (often sociologists and anthropologists) see ethics policies as a veil for power and economic exploitation.[3] I think both views have some merit to them, and yet I subscribe to neither.[4] It is naïve to assume that a policy is a safeguard against the system producing it, and policymaking does not follow special rules just because it revolves around something called "ethics." Conversely, it is arrogant and foolish to suggest that everybody doing "ethics" is part of a strategic conspiracy, as some critics seem to imply. Furthermore, both positions seem to share with market thinking the view that peoples' moral engagement and some kind of "market forces" operate in totally different and unrelated registers. We need to understand the power effect of ethics policies without bifurcating the agency going into "ethics" or "business" and without reducing moral agency to strategic calculative agency of the type suggested in crude versions of market thinking.

As already stated, I do not presume that what people do for nonselfish reasons necessarily brings about something that will be agreed upon by all the involved (human and nonhuman) actors as "good." Moral agency can be cruel. Ethics as an academic discipline and as a policy arena is nevertheless usually expected to deliver coherent answers to the question of what to do if one wishes to sidestep self-interest and thereby pursue something positioned as a "common good." The notion of "really, really good"—irrespective of your point of view—is an important source of motivation, and it is part of creating visions of a world which is yet to come. It has performative effects for the present to imagine what is not yet in existence, but as pointed out by Carl Elliott, there is no external point outside the world from which we can generate a truly coherent moral world. Ethics as a discipline has no straightforward programmatic effect, not even on moral agency.[5] If we want to understand what ethics policies do, we must acknowledge their performative effects without assuming the policies to be strategic either as a program for a better world or as a calculated deceit.

More specifically, I think we should pay more attention to the performativity of the moral paradigm of market thinking as it names and frames the problems policies address. We must see what the inclination to keep persons and commodities apart *does* to the exchange of ubjects. My suggestion is that it focuses the attention on manageable problems and that it produces areas of what Murray Last called *not knowing*. Last suggested that "under certain conditions not-knowing or not-caring-to-know can become institutionalised as part of a medical culture" and that by investigating what people prefer not knowing, we learn about what they value.[6] There is a conspicuous tendency among many observers (especially market opponents) to imagine that ubject procurement involves no profit-making, despite the fact that many for-profit companies are involved in it.[7] Such companies operate within the affordances of "processing fees" and "compensation," and I take the quite outstanding ability to ignore their profit-making as indicative of an interesting form of moral agency exerted in the form of not knowing. To understand ubject exchange in practice, we need to engage what exchange partners do not know and what facilitates this dedicated, and yet tacit, not knowing.

In the first of my two cases, the one focused on blood, the moral paradigm relegates monetary concerns to a realm of information to keep blood as a material entity beyond commoditization. Blood is not commoditized, but knowledge extracted from it is capitalized upon through patents. The moral paradigm facilitates this process of bifurcation by fixating the sense of ethical gravity on blood as a substance. In the second case, where bone is desired for its material properties, the monetary concerns cannot be relegated to a patent realm. As in organ procurement organizations, however, the hospitals collecting and shipping bone can charge "handling and processing fees" and ask for "compensation of expenses." The moral paradigm here does its work by making the calculation of expenses an embarrassing topic that the involved persons do not interrogate. The result is a room for profit through overcharging of "compensation." Both modes of "obscuration" of the price-setting element of the exchange are related to concerns about commoditization of the human body. Following the case studies, I first compare my Scandinavian findings with some oft-cited survey results and then place them in a context of other studies and cases in order to tease out what they might say at a more general level about the role of money in ubject exchange.

(Ex)changing Blood for Genetic Research

In the late 1990s, a group of Swedish researchers carefully studied the news about an Icelandic genetic adventure. An Icelandic born Harvard professor by the name Kári Stefánsson proposed the construction of three databases in his home country to be managed by a company registered in the USA, deCODE genetics. He took the position as director, the parliament passed legislation to facilitate the operation, and a Swiss pharmaceutical company supplied ample venture capital. This was the beginning of one of the most impressive biotech adventures ever in any of the Nordic countries.[8] deCODE constructed a genealogical database and began constructing a biobank containing blood samples from consenting donors, and began also work to construct an electronic health sector database of all medical records from 1915 onward.[9] The latter was never completed, partly due to international opposition to its use of presumed consent.[10] Many internal observers assumed that deCODE would use presumed consent for genetic research and did not realize the difference between the three databases. By almost any standard, however, the company was a huge *scientific* success, and it quickly published numerous papers in *Science* and *Nature*. The group of Swedish researchers paying especially close attention did so because they already had a lot of the material that Iceland was now beginning to collect in the form of a population-based biobank called Medical Biobank in Västerbotten County in northern Sweden. Not surprisingly, they wanted a share of the venture capital and, by extension, the scientific progress.

Entering the Field

Medical Biobank was founded in 1985, and by the time of deCODE's establishment, it contained blood samples and medical information from the majority of the county's adult population. The biobank is located at the university hospital in the town of Umeå, the largest town in the region with approximately 105,000 inhabitants. I got to know the group of researchers behind the biobank as I went to Västerbotten in 2001. In the course of the following 3 years, I conducted 12 months of fieldwork moving back and forth between donors, venture capitalists, politicians, and health professionals collecting, storing, and distributing blood as well as university and industrial researchers using the blood. The blood and data for Medical Biobank had been collected during medical examinations offered to all county inhabitants at the ages of 40, 50, and 60 years at 36 public health-care centers in Västerbotten. People attending an examination spend two and a half hours at the health-care center. They see a nurse, get some blood tests done, are invited to donate 20 ml of blood for research, and fill in an extensive questionnaire. They have a conversation with the nurse based on their questionnaire and the test results, and during this conversation, a "risk profile" is produced. The questionnaire is stored for research along with the donated blood. I interviewed 57 inhabitants from 5 different locations in a 90-min break they had while waiting for test results, and I conducted focus group and individual interviews with the staff responsible for the medical examination and collection of blood. But most of my time was spent hanging out with competing researchers in the company and at the university, with biobank staff and with relevant politicians involved in regulating access to this research resource.[11]

Västerbotten is a region of about the same population size as Iceland and with some of the same genetic characteristics in terms of late settlements, relative isolation, and known rare diseases running in particular families. With this biobank already operating, Medical Biobank considered itself in a good position to compete with deCODE. To get things going, a start-up biotech company, UmanGenomics, was given all commercial rights to research on the biobank material. The blood stored in the basements of the local university hospital was now referred to by local researchers and politicians as a *gold mine*, and the most enthusiastic researchers said it was *unethical* not to explore its financial and scientific possibilities. But the controversy resulting from the international misunderstanding about the use of presumed consent in Iceland was also daunting. To avoid similar complaints, the managing director of UmanGenomics initiated work on an ethics policy, and when finally the company was launched, it was with international appraisal of this policy in *Nature* and *Science*.[12] I focused my study on the role of ethics in regulating access to genetic material. Because ethics was framed as a matter of "policy," I construed my field as three interrelated levels of policy work: (1) I tried to understand how an *elite* level (what I called *policymakers*) named and framed the "ethics policy," (2) I followed the policy through its translations among the *nurses* who (as *policy workers*) were working with the policy of informed consent, and (3) to the *donors* of biobank material (the *target group*) for whom (and on whom) the policy was supposed to work.

The ethics policy stipulated the use of informed consent, oversight by an ethics committee, and public oversight through partial public ownership. The idea of public ownership was contested because it turned out to be at odds with laws governing public agencies, and later also at odds with moods governing venture capitalists, and in the end, it was abandoned at the request of the company director to facilitate better commercial appeal. On the surface of it, the ethics policy had little effect other than, perhaps, making informed consent forms longer, and it was in fact described by the director mainly as an attempt of dressing up existing procedures as a "policy" to communicate clearly to the general public that the company's endeavor was unproblematic.

At a first glance, this approach to ethics appeared to be exactly what Nancy Scheper-Hughes describes as ethics as veil of exploitation "in which human bodies are the token of economic exchanges that are often masked as something else— love, altruism, pleasure, kindness."[13] Parts of human beings had been collected and exchanged and were now used to facilitate financial gain with—it appeared— some talk of ethics as mere window dressing. There was something about that narrative that kept nagging me, however, and throughout my fieldwork I returned to this easy conclusion to contemplate its merits. The employees at UmanGenomics were very engaged in a range of ethical issues, and they were very strict on not referring to the blood samples without respect for people's donations. Blood was not to be viewed as a commodity, they said, and they really seemed to mean it. Furthermore, Swedish authorities were developing new legal frameworks to prohibit commoditization of human tissue. Moral concern and ethical rhetoric thus worked toward excluding tissue samples from commercial exchange even if the specific ethics policy was just a sort of epiphenomenon in this process. Even more puzzling, it seemed that the stored blood which had been lying in the basement for years without causing any concern was, rather than being degraded and objectified, increasingly *ascribed* subjecthood through the new commercial arrangement. In fact, concerns about monetary incentives for blood donation were a relatively new phenomenon in Sweden and somehow related to the rise of (commercially interesting) biobanking.

Selling Blood?

When in 1970 Richard Titmuss published *The Gift Relationship*, he briefly mentioned that Sweden, in contrast to his overall argument, had a relatively well-functioning transfusion system, which nevertheless relied 100% on monetary incentives. Titmuss probably misinterpreted the role of financial incentives in Sweden. Swedish blood exchange should be understood in the context of the state-centered Swedish health-care system. The project of constructing a welfare state providing health services and housing to everybody began in the late nineteenth century and found its impetus in particular in the workers movement, the temperance movement, and the free churches. In these movements, there was a notion of being mutually obliged, and the demands for better services were typically countered by demands for self-discipline and suppression of personal idleness. Technologies of state were coproduced with technologies of self. State and citizen became engaged in a directly obliging relation characterized by the ambition of fairness.[14] When the public health-care system needed blood for transfusion purposes, citizens were in a sense obliged to donate, not legally but morally. However, as taking part of the day off from work to go to a blood bank would incur a loss of income on the individual, a system of financial compensation was decided upon. Money was not used to incentivize, but to give a fair compensation.

Even if Titmuss underestimated the sociocultural and historical context of the blood donation system, it was nevertheless correct that blood and money changed hands, and when the system was debated, focus was on the practical aspects of ensuring a sufficient and safe supply, rather than issues of commoditization.[15] Monetary compensation was not so much a market issue as a fairness issue in the state/citizen relationship. In conjunction with the rise in transplantation technology, trade in human bodies gradually became coined as an ethical problem in its own right, however, and policies were enacted to keep ubjects out of market reach. Nevertheless, in the transplantation law from 1995, blood, hair, breast milk, and teeth were explicitly mentioned as exemptions from a commercial ban.[16] Paul Rabinow finds similar French laws somewhat arbitrary in its demarcation of what can legitimately be sold,[17] but such rules did not necessarily appear arbitrary or irrational to the Swedish donors. One morning, I was interviewing a man who had just turned 40 and who considered the medical examination during which he had donated blood to UmanGenomics "a kind of a birthday present." He was wearing a fine new red T-shirt, looking both casual and cheerful but also slightly anxious about the results of the sugar level test that he was about to receive. He argued that "before we knew about DNA," it had been appropriate to distinguish between renewable "goods," like blood, hair, teeth and breast milk, and parts of the human body that cannot be extracted without the donor suffering an irreplaceable loss. With the discussion from the previous chapter in mind, we could also say that the sense of control was not as heavily challenged with replaceable ubjects. Interestingly, this man, like most of my informants, thought that today blood should no longer be bought and sold, and he explicitly linked this ban to new meanings attributed to DNA created by way of genetic technology. A new axis of truth in the Foucauldian sense became the basis for doing ubjects differently: even blood could now remain partially connected to donors after leaving their bodies.

I will go further into donors' attitudes toward the selling of their blood. First, however, it is important to observe that it was not until January 2003, when the Biobank Act was enacted by the government, that it became illegal to sell blood in Sweden.[18] From having had a system of blood donation for transfusion purposes using economic compensation to donors, Sweden has thus withdrawn blood from exchanges involving money. The attention of regulators has become firmly fixed on ensuring that blood as a *substance* is not bought and sold. Clearly, there are many economic aspects of the transfusion industry (as well as biobank research), but monetary incentives to donors are no longer legitimate, and policy initiatives are enacted to decommoditize blood. Only a handful out of the 57 potential donors that I interviewed found it legitimate to consider receiving money in exchange for their blood. Importantly, even this little group of people can be said to hold a decommoditized view of their blood, as discussed in more detail below. In some respect, this reluctance toward monetary incentives contradicts the presumption in market thinking of rational actors always seeking to optimize their economic stakes in the world, and it is remarkable considering that many of them previously had received money for blood for transfusion. As I now turn to some of their reasons and ways of positioning themselves, it is important to remember that unlike donors of blood for transfusion purposes, research participants are being asked for access to an aspect of their person. The blood is thereby personified through the very positioning of donors as research participants. And unlike state procurement organizations for blood, commercial biotech companies have a hard time pretending to offer money for the sake of fairness only.

Wishes, Motives, and Uncertainty

I asked all the people attending a medical examination whether they would be willing to sell their blood. It should be kept in mind that most of them just had donated 20 ml for free; hence, it was a hypothetical question. In response, some just stated that it would feel wrong without specifying why, as this woman exemplifies:

> I don't know—I guess it would feel wrong somehow.... No, I prefer giving it. It feels more proper than selling, eh, I don't know. I guess I've never really thought about it.

Others tried to give reasons or specified their feelings. These reasons and specifications usually clustered around three partly overlapping themes that I have categorized as wishes, motives, and uncertainty.

Concerning *wishes*, many donors stated that it was more important for them to see research thrive than to receive a few pennies. Some argued that the cost of research should not be increased by anything unnecessary like compensations for blood samples, as the cost of good research was already high. One woman, whom I interviewed at a health-care center in the town of Umeå, said:

> I've never really thought about [whether I would sell my blood].... For my part, it would feel better if what you received in return was less suffering in the world, rather than—no, I don't feel like being remunerated.

Another woman in the interior region in a similar vein explained:

> What's supposed to be for the benefit of the future that—that you cannot charge them for. It is so little anyway [20 ml].... If it's to be of any use, it shouldn't be too costly.

These people articulate a vision of research for the benefit of mankind and wish more research rather than money for themselves.

Reasons relating to *motives* reflect what people expect of themselves and others when money is involved. Some people state that they would feel their own motives misunderstood if money was involved in their donation. One man thought it would be embarrassing to receive money, as people would think that he was really desperate. Several people said that it would be better to donate without payment because then it would be more voluntary—implying that in commercial exchanges, people would act out of financial necessity, almost as reminiscent of Titmuss' argument and a point recurrently discussed in the debate about monetary incentives for ubject donation. Finally, some feared that it would make people donate for the wrong reasons, thus donating research material of a lower quality (again similar to arguments posed by Titmuss).

Other interviewees were more concerned about issues of *uncertainty* and control. Several people argued that if they had sold their blood, they would have no control over it anymore. One woman had not wished to donate a sample and explained:

> I guess the worry is where it might end up. What it's gonna be used for, if it's only the stuff that stood in the letter or if it's—well, if it's circulated further—if it'll be resold, sent along to something.

Umeå is by most international standards a relatively small town, but from the perspective of this woman, it was a big industrial center pretty far away. She was living 5 h by car northwest of Umeå where the seemingly endless forest gradually gives way to mountains. She did not give the impression of living on the margins of anything, however, but of feeling closely related to everyone around her. Her statement illustrates the difference between the face-to-face relations in which she lives her life and those mediated through impersonal trade. It was typical of many donors, also those in Umeå: in commercial relationships, the buying company is thought to be free to do whatever it wants with what it has *bought* (see Chap. 2 on the rise of the notion that one should be entitled to do what one can afford to do). Private research in general was often related to notions of uncertainty. When asking my informants whether they saw any difference between private research and research undertaken by a university—which is in Sweden generally understood as public research—it was often stated that with private research, "you could never be 100% certain." Hardly any informants could specify what types of issues they wanted to ascertain themselves about. A few were concerned that the "money devil" would take control; one person thought this had already happened. Others were convinced that a private research industry could be just as good as any public institution, perhaps even better. However, they often went straight on to mention the rule of law and other safeguarding rules of the Swedish state as a guarantee that private research would not do anything wrong, thus again framing the issue of industrial research in terms of levels of control.

The anxiety provoked by the uncertainty associated with private research may help to explain why people do not opt for the money they could have made, if selling their blood was still legal. Money seems a poor substitute for public authorities assuming responsibility, and donating your blood freely as a gift for beneficial research seems a better instrument for making public authorities feel obliged to ensure the interests of donors than money from a private company. These participants in genetic research seem to opt for a "gift relationship" in as far as they want the users of the blood to feel certain obligations.

Even if people care about what type of relationship they engage in with their donation, their interest in UmanGenomics as such is limited. The company was generally seen to be an agreeable venture. However, during the spring of 2003, an internal conflict among different researchers and fractions of the university about the entitlements to the biobank material was thoroughly exposed in the newspapers, and at that point some donors said the stories had put them in doubt.[19] Few donors had engaged the issue with any particular interest, though. When asked during this period what they had heard about the biobank and company, a typical response was:

> Nothing more than I've read a bit, the headlines in the newspapers, I haven't gone any deeper into it.... I think there's a bit of politics in it today, so... [timid laugh] that's why I avoid going into it.

Medical research is not supposed to revolve around money and politics. Also in 2003, some nurses told me that they experienced people abstaining from donations with the explanation that this company made the objective of research seem uncertain. A few nurses thought that the commercial interests in the blood they collect were an infringement of the trust they had established with their patients, and they simply decided not to ask patients for samples anymore.

The "Right to Sell"

In genetic research, the donated blood is understood as demanding a form of trust which is not expected to emerge from a commercial exchange. Therefore, most people opt for other forms of exchange. Thus, they demarcate their bodily resources as different from other resources and embrace a decommoditized attitude to blood. Despite the prevalence of these responses, a few people stated that they would not mind receiving money in return for their blood, as mentioned above. However, even these people described exchange of (what I call) ubjects as different from other exchanges. One man laughed as if I had told a dirty joke when I asked if he would have liked to receive money himself. He nodded and showed with his hand how he would let the money slide into his pocket, but raising his eye brows, he remarked also that it would have to be "big money." Two people said it ought to be legal to receive money because people are "actually doing a piece of work." They would receive money for labor, not blood. Only two or three people (one seemed to change position in the course of the interview) were of the explicit observation that they should be free to sell their blood. One of them, a very muscular ex-soldier with a slightly confrontational attitude, explained:

> Well, we've got free trade!... I can sell my kidney if I want to, today!

I remarked that it happened to be illegal in Sweden, and he responded:

> It's illegal in many countries, but it's not unusual around the world and Europe to sell them. And you can get them in illegal ways too, but that's not the point. The point is who owns my body? And that's *me* who does that, eh?! It's not the health authorities who decide over it, it's I, and therefore, I can do as I please!

He explained that it was only fair to receive money when you contribute to a project where you do not share the interest of the researcher—such as surveys executed by, for example, financial corporations. However, health-care-related research was special, he went on, because here you share the interest of the researcher, and personally he would not like money for participation in medical research projects. So blood was different after all. Intriguingly, however, he had decided not to donate despite this observation:

> It's a matter of safety issues and handing out information and all that. I've worked with security issues [in the military], so I know what it's about.... If you want a piece of information, you'll get it, eh! No matter if you're entitled to it or not. That's how it works.

His reasoning is similar to that of the majority in that it hinges on the same notions of public control, healthcare being an exemption to other areas of trade, and him being personally uninterested in selling any part of his body. It departs from the general trend in not trusting public authorities and in thinking of his body as safeguarded in the best manner through a proprietary relationship with himself as commercial owner. Often, when I have been participating in conferences debating property rights to ubjects, I have been struck by the similarities between the reasoning of many academic market proponents and this particular man, in that property rights are seen as a sort of safeguard against both "state" and "market," rather than as a starting point for enhanced ubject flows. A right to sell is propagated to *defend* the body (making blood into a part of it even after it has left the veins), *not* because the body is seen as a mere thing.

Ethics as Facilitation

One could read the analysis above as mainly criticizing the perception of ethics as a veil for commoditization. To make clear that this should not be seen as indirect support of the notion of ethics as a safeguard against the market, I will now turn to a critique of this second set of ideas circulating in the literature. The ethics policies and the moral reasoning subtracting blood from market exchange provide not only constraints to, or a defense against, "market forces": they are an essential part of commercial research conditions on positive terms. The point is that ethical rigor to keep blood out of the market is part of facilitating genetic research as well as part of creating a move toward a new consumer-friendly distribution of entitlements by relegating profit from blood as a substance that donors may share in, to a realm of research results accessible only through intellectual property rights. The ethics policies in this field construe a particular form of problem that is compatible with commercial interests. Attention is directed at problems that are, in a sense, manageable. In particular, it is striking that the concern with commoditization of human tissue has left very little concern for amplified capitalization of knowledge.

During the work with the ethics policy presented by UmanGenomics, it was decided that the company and the biobank should be completely different entities to make it clear that a company could not own blood. Secondly, as the Biobank Act made trade in blood illegal, a profit-related financial compensation of the public biobank originally intended was deemed unethical (and at odds with the new law). It illustrates how it gradually became a key concern for policymakers to ensure that blood was not bought and sold. The company representatives explained this to me several times: "We're not selling genes, you know. We are in the research business." County officials also explained that not a single drop of blood would be sold. That research results, "research business," can be bought and sold is less debated.

The trade in knowledge depends on intellectual property rights (IPR). This type of entitlement has a social history of its own that I will not explore further here.[20] What I want to remark is that many donors do not acknowledge intellectual property claims with respect to patenting of genes—in spite of its support in law and ethics policies.[21] The people behind UmanGenomics' ethics policy have nevertheless used IPR in an attempt to install an ethically legitimate form of market exchange that divides substance and knowledge and excludes the former from trade while amplifying the value of the latter. Donors may hold no commercial rights in their own body, and potential profit remains safely in the hands of the knowledge industry. As a benign side effect, negotiations of *price* can now be done freely as matters of patent deals and in accordance with the ideal of the free market. I am not suggesting a conspiracy, but I do imply that policymakers tend to have a pragmatic feeling for the issues that can be successfully addressed and that gene patents are not deemed one of them. The moral focus on blood as a substance gains strength from the cultural narrative of material continuity between body and person and relegates concern to a (financially) relatively harmless realm.

The control of knowledge is of much greater importance to the way inequality works in the research intensive so-called bioeconomy than control of substances. The wealthy nations thrive on developments in the global terms of trade which has for more than 40 years consistently favored those in control of technological knowledge, rather than those delivering raw materials. The global enforcement of intellectual property rights (IPR) has been at the center of this development. They are also central to current disputes of third world access to medicines and other patented health products. The rapid expansion in IPR has been concomitant with the concern with commoditization in the wealthy nations, but while laws have been changed to ensure that people are not trading in human substances, the European Union has agreed to demands made by the United States that extensive areas of intellectual work can be patented, owned, bought, and sold—and kept out of reach for those who cannot afford it. Inequality thus thrives on economies that maintain a sharp distinction between substance and knowledge about that substance; and reserve the notion of ethical problems to the handling of substances.[22]

If ethics is not a veil for "real" commoditization of blood as a substance, it might still be seen as facilitating trade in research results on market terms. It masks the interconnectedness of two exchange regimes: the local trust-based exchange regime for tissue feeding on the state/citizen obligations versus the international commercial exchange regime of bioscience.[23] Successful commercialization implies skilled maneuvering in a landscape of moral intuition. As Sarah Franklin notes, "Successful capitalization … is all about recognizing the links between the hype, the product,

the market, the science, the government, and the goodwill of the general public."[24]
What we see in Sweden is an example of skilful maneuvering in the moral landscape by the people presented with the task of enhancing the public/private interaction in this field to make business out of tissue collections. Rather than having had
a planned strategy, they seem to have responded well to contingent ethico-political
signals. They have articulated latent concerns in ways which have temporarily
fixated the debate on blood as a substance and individual rights to informed consent.
This fixation contributes to the separation of a national realm of citizen rights and
duties from a global de-personified data exchange system where money may establish and split up connections undisturbed by notions of mutual obligation. The tools
in this process presuppose a particular national history and have included legal
amendments and organizational setups designed to signal political will to protect the
citizen while in turn delineating the protection to a matter of providing the individual
with the right to informed consent. The potential changes related to privatization of
medical research in this way avoid articulation and public contemplation. Moral
reasoning conditions monetary gain and vice versa.

(Ex)changing Hips From and For Implant Therapy

If the biotech industry's capitalization on blood is facilitated through bifurcation
of substance and information in a paradoxical move through which blood as a
substance is simultaneously reinterpreted as the essence of the person and beyond
trade while the informational aspects of this "essence" can be patented and sold,
then how may one capitalize upon ubjects desired for their material qualities? To
address this question, I now move on to the study of hips. Again I suggest that we
might better grasp *how* exchange systems deal with ubjects if we do not presume
that the moral paradigm of market thinking and its increasing number of legal
manifestations merely mask what is really a "market." We need to understand
what the moral ideal *does* to the mode of exchange while simultaneously acknowledging that ubject exchange has genuine monetary implications that cannot be
discursively erased.

The hip is a fascinating ubject and relevant for the study of the interface between
the human body and the commodity domain for several reasons. First and foremost,
it comes in two versions. One is made of metal, also known as a "prosthetic device."
Hip replacements, or total hip arthroplasties, are with some international variation
among the most common types of elective surgery in most wealthy nations.[25] The
other version of the hip is made of bone and grown inside the human body. This type
is also a desired and readily exchanged ubject used in transplant surgery, although its
exchange is not as well known as that of the prosthesis. Successful bone transplants
have been conducted since the second half of the nineteenth century and are among
the most common types of transplant.[26] Bone can be procured from cadavers and
from patients undergoing a prosthetic bone replacement, primarily hip replacements.
In the UK, it is estimated that bone is used in approximately 7,500 operations a year.

This should be compared to an annual rate of 3,000 organ and 2,100 hematopoietic stem cell (bone marrow) transplants. In the USA, in 2006, the number of bone and tendon transplants was estimated at 1.5 million with approximately 22,000 annual donors,[27] while in Denmark, approximately 4,000 femoral heads are used in around 1,500 patients a year. As indicated just by the quantity, there is reason for some observers to refer to bone grafting as a billion dollar industry.[28] Practically all bone in Denmark is donated by people receiving a hip prosthesis. Recently, the exchange systems for bone in both the EU and the USA have been restructured to optimize safety.[29] With huge amounts at stake, with both a metal and a bone version, and with recent attempts of restructuring the exchanges, there are many good reasons to choose the (otherwise unfashionable) hip transactions to explore how the monetary aspects are handled at the microlevel when the exchanged ubject is not supposed to have a price and cannot be construed as a purely informational asset.

As in the Swedish case, I focused my study of hip bone on a policy. I took point of departure in the EU Tissues and Cells Directive meant to regulate bone procurement, storage, and exchange. I collected documents and interviewed *policymakers* influencing the Danish translation and implementation of the Directive, and I observed and interviewed *policy workers* (surgeons collecting bone and bone bankers storing and distributing it in light of the new rules) and members of the proclaimed *target group* (donors and recipients). In this study, I focused on what I have called policy workers and, in particular, on what happens in the relation between bone bankers and surgeons. There are 23 registered bone banks in Denmark, but actually, there are only 17 organizational units because some of them are run by the same people with different sites of storage each having their own license. One or more representatives of all 17 were interviewed at least once, and their views and experiences were mirrored in material gathered through interviews with policymakers, surgeons, donors and recipients, and information from participant observation at one hospital. In tandem with this, I visited crematoria and interviewed people working with recycling of prosthetic metal, and I invited also the bone donors to reflect on the potential recycling of their new hip.[30] In the following, I give an overall description of the form that hip exchanges take, and I organize this description in terms of steps of disentanglement and subsequent exchange.

Bone on the Move: From Waste to Value

For a hip bone to become exchanged, the bone must be disentangled from the body. When and how is this done? First, the right bone must be identified. It is not every type of bone that can be transplanted. The agency of the bone must be carefully assessed to ensure proper capacities and absence of unwanted contamination:[31] bone excised from operations on indication of fractures, osteoporosis, or cancer is not supposed to be used. However, there are no major issues of tissue compatibility as with organs, and this makes bone into a relatively mobile ubject. Nevertheless, as Michel Callon points out, for an object to travel, "it is necessary to cut ties between the thing and the other objects or human beings one by one. It must be decontextualized, dissociated and detached."[32] This goes for ubjects too. So when and where does the disentanglement of the hip bone take place and who provides the bone?

Since, in Denmark, bone used for allogenic transplants (transplants between different persons) comes almost exclusively from patients undergoing hip replacement operations, the patients are typically asked about bone donation during the examinations determining whether they should undergo the operation.[33] I observed these conversations at a major hospital in the vicinity of Copenhagen. Ironically, for this type of surgical ward, I had a broken leg when I first began these observations, and with my crutches and leg in plaster I looked more like a patient than an observer, at least during the first couple of weeks. Nevertheless, when observing the conversations I would try to withdraw into a corner and interfere as little as possible. The surgeon and patient typically sit in front of an x-ray showing the depleted cartilage, and the surgeon points to the bone and explains:

> This is where the problem is; the joint doesn't work, and this [pointing at the femoral head] we take out. You can't use this. But we can. Is it alright if we keep it?

From my position in the corner and in my project, these conversations featured as "informed consent for donation." I gradually realized, however, that for most patients, this had little to do with bone donation and bone banking. It revolved around understanding a forthcoming operation often causing a lot of anxiety. I heard no questions posed concerning the donation as such, and the interviewed surgeons rarely experienced patients declining donation. Sometimes, however, they would forget to ask and therefore had to get a consent signature after the operation.

Why would a patient want to donate a bone? Ironically, the first step in creating exchange value out of the bone is to designate it as waste as indicated in the above conversation ("You can't use this. But we can."). The image of *waste* is productive in facilitating exchange as discussed in Chap. 3. Donors embrace this image and highlight the possibility of *helping* others with something they cannot use themselves. One man interviewed at his bedside in his white hospital dressing gown said: "Really, for me it's just a piece of waste, right, so I really don't care. If it can be put to use, it's fine." And an elderly woman, whom I interviewed in an office on the hospital ward shortly after her operation and while she was still looking a little fragile, explained:

> Well, eh, I thought it was quite all right. I just said as long as you don't take any more than what you're supposed to, it's quite all right [timid laugh]. Really it doesn't matter 'cause it's thrown in the bin, right?

Two men drew reference to potential alternative uses of the bone, which were deemed undesirable, by saying either "I hadn't expected to take it home to the dog" or "I'm not going to use it for anything—I haven't got a dog, you know." Both men were joking of course and accompanied their comments with shrewd smiles. When in various contexts I have recounted this image of a dog lying in the living room with a hip, people have typically reacted with a kind of grotesque laughter. It is simply too weird to accept as a plausible use of a good piece of bone, and yet no obvious explanation seems to be at hand when people seek to explain why. The fact that it appears unfit for animal consumption shows that the bone is not just any kind of waste and that it is not *just* an object. It remains partially connected to the subject it once inhabited, or, rather, the wrong form of usage reconnects it to the "donor." Waste is an in-between state of recategorization, a state in which values transmute but do not disappear.[34] The categorization of bone as waste does not capture the meaning of the bone as such; rather donors and doctors use the categorization to establish a shared understanding of the implications of a donation: the donor does not stand to lose anything by letting go of the bone. It can be disentangled without the donor losing control. For the bone to be disentangled, this sense of certainty must be established, and with it the meaning and value of the bone can remain relatively undefined.

How Bone Acquires a Price

It is not enough to get the bone disentangled from the donor to make it travel. To understand the complexity of bone exchanges, one must appreciate the organizational context and the various steps from provider to bone user.[35] Thirty years ago, Danish surgeons would themselves procure the bone they needed and keep it in a freezer at the ward. Today, however, the health-care sector is differently structured, and most of the bone is used in hospitals or wards other than where it is procured: the uncomplicated hip replacements mostly take place in regional hospitals with lower levels of specialization, while the more complicated surgery, during which bone is used, mostly takes place in specialized university hospitals.[36] Bone banks coordinate the distribution of the bone. If from a legal perspective the donor is the initial provider, from the perspective of the bone bank, the important provider is the *surgeon* because it is the surgeon at the low-tech unit who must be motivated to collect the bone used for more advanced therapies offered at university hospitals.

All hospital units are asked to pay a certain amount for each femoral head they request. According to market thinking, price setting is supposed to *motivate* work, but in this case, price setting seems to *produce* work motivated by other agendas. In the early system, where the individual hospital units kept bone on store until it could be released to the next patient, this was not so. However, in tandem with specialization and differentiation of tasks between different units, principles from New Public Management (NPM) have been introduced into the health sector. NPM also embodies elements of market thinking and involves financial accounting for each piece of work based on the assumption that direct remuneration stimulates efficiency. Therefore, the shipping of bone from low-tech local hospitals to regional bone banks and further onto university hospitals involves a monetary accounting problem. Few surgeons collecting bone are particularly aware of this problem and how money changes hands. They are primarily motivated to collect bone by being told of its clinical utility; money is not used for this purpose. All the same, when each unit is run according to individual budgets, the shipping of a package of bone necessitates the calculation of the costs to be recovered. How, then, is this amount determined?

Bone-bank managers usually talk freely about the bone as having a "price" and usually state this price in a matter-of-fact way when asked. The amounts stated varied between 1,500 and 4,500 DKK, partly depending on whether the bank had begun using the new and more expensive tests as a result of the new safety regulations introduced with the EU Tissues and Cells Directive. Bone bankers take care to explain that the price "just covers the costs." But who determines what are relevant costs, and why do different bone banks have so different costs? In most instances, it appears that the "cost" is the price paid in the laboratories for the various tests. This was particularly evident in relation to the new and more expensive way of testing for HIV mentioned above. Most bone banks have simply added the price for one test (1,400 DKK) to the price of each femoral head—or they planned to do so. Some claimed to add the price of the container in which, according to the new regulations, the bone must be transported. One bone-bank manager took for granted that the cost of freezers and working hours were calculated into the price, and another said that even the working hours of the surgeons had been included. However, the inclusion of these extra costs in these two bone banks was contradicted by the prices being lower in these banks than in bone banks claiming to make no such assessments. They could not specify the price of the various elements mentioned. In addition, bone banks necessitate physical workplaces, cleaning and maintenance, none of which were calculated into the price. One bone bank stated two different prices for the bone it delivered, 3,500 and 4,000 DKK. It turned out, however, that taking 4,000 DKK was an embarrassing mistake when one staff member had forgotten the correct price. The fact that the higher price had caused embarrassment illustrates that price setting is not a simple pursuit of maximum profit (as market thinking would suggest) and that it follows its own rules: it is not strictly related to expenses; it is not a function of demand.

Not only does the price not take into account all the adjoining costs of bone, it also fails to take into account all the bones which must be disposed of either because they are tested positive or because patients do not reappear for any necessary additional testing. When recovering only the cost of the tests for each *successful* bone transfer, the bone bank will have to cover the cost of all the discharged bones without any compensation. One bone banker admitted that the price was "frankly, just a rough estimate," and several others downplayed the relevance of the price explaining, for example, that "it is not that important as long as it all remains within the public authorities." At another bone bank, they admitted that the price had simply been copied from the price another bank was taking. Nevertheless, it is stated again and again that the price reflects the cost; that the income balances the expense. It seems, however, that "prices" serve a different purpose than making bone banking into a self-reliant and cost-neutral activity. The notion of cost neutrality and the wish to avoid making bone transfers into what could be described as a business facilitates a relatively relaxed attitude to the calculation of expenses—as long as each organizational unit can signal financial responsibility upward in the bureaucratic system by claiming to have made a calculation.

A representative of the authorities was aware of the ambiguities concerning price setting and described this very ambiguity as part of "the oil-like disposition that keeps the machinery going." He pointed out that complex organizations have many overlapping tasks, using its buildings and facilities for multiple purposes and performing not only bone transfers but also policymaking, documentation, maintenance, development, and research. As a consequence, it is simply impossible to determine the exact cost of every bone. What is striking is the paradoxical coexistence of this insight and its adjoining casual attitude to price setting (or we might say dedicated not knowing), on the one hand, and then the strictness of the claim that prices cover only the costs, on the other. If Michel Callon points to the performativity of the economic discipline in shaping exchanges expected to take the form of a "market," we might add that the moral reasoning surrounding ubjects that are *not* supposed to be traded as commodities has performative effects for the calculation of recovery costs. It displaces certain types of reasoning or makes them appear irrelevant. Other types of reasoning associated with New Public Management (NPM) nonetheless necessitate balance sheets and audits to keep track of expenses and demonstrate responsible conduct. But the setting of a price hardly delivers the benefits that proponents of market thinking tend to associate with price setting according to supply and demand mechanisms. Several other ways of calculating the price could be imagined: not only could other expenses be included, also the unit of bone could be different. Rather than counting the number of femoral heads, the quality and weight of bones could go into the price, which would be relevant if price should reflect demand.[37] But what we are observing is essentially a different type of exchange system. In the following, I expand upon the languages used to describe bone transfers beginning with patients and bone bankers to unfold further the specificities of the exchange system for this ubject.

Discourses of Bone Transfer

Just like donors of blood in Sweden, the Danish bone donors state clearly that they are not interested in receiving money for their bone, and both donors and recipients recounted—with disgust—stories about people in low-income countries selling their kidneys. One of the donors I interviewed at the hospital a few days after his operation explained: "Well, I've asked to be rid of it, and so on, I've had an improvement, so it would feel ridiculous [walking off with money]." Another man explained that though he would not personally like to receive any money, he did not care if the bone bank sold the bone to somebody else as long as the national health services took care of the trade. Others found all commercial transfers worrisome and compared it with "organ trade." Some added, however, that it might be necessary to compensate a hospital for its expenses associated with collecting the bone. One such woman added: "as long as it doesn't turn into a business plan—so that they pinch something I might have needed anyway." Note how acceptance of monetary aspects of ubject exchange is balanced with concern about a potential loss of control: a phenomenological category of body is coproduced with the expectations people have to exchange forms, and just like blood donors in Sweden describe commercial exchange as related to losing control, these bone donors worry about what commercial incentives would imply for their bodily control.

The image of the public authorities "all drawing from the same account" was typically referred to as a reason for not thinking about money at all. When reminded of the private hospitals, it was said that *compensation* of their expenses would be in place, but not profit. Again the notion of compensation as clearly distinguishable from profit comes forth. Apparently, parts of human bodies are, from the perspective of the people donating and receiving them, not supposed to be transferred *because* of money; nevertheless, economic aspects of the transfer must be dealt with, but preferably in ways not influencing the motivation of donor, recipient, or surgeon.

If donors generally talk about *one* bone, *my* bone, or *the* bone, indicating a specific unit of bone, surgeons and bone bankers tend to use the generic term *bone* without an article. In general, the language used among surgeons and bone bankers is much closer to what would be expected in a business dealing with unequivocally nonhuman materials. The common lingo among surgeons is to talk about procurement from donors as *harvesting* and the subsequent transfers via the bone bank as *buying* and *selling*. This can be compared to a farming terminology. While these expressions were used freely in interviews, I did not hear this terminology used in front of patients. Bone bankers also drew on the farming terminology but in addition talked about who *produces* or *delivers* and who *consumes* bone, using what more seems to represent something closer to a manufacturing terminology. When looking more closely at the way bone bankers reason about the transfers, it remains obvious that "buying and selling" in this field operates very differently from what adherents to market thinking would expect, as already indicated in relation to the setting of a price, and yet there is something consistent in the way greater distance from donor bodies can be detected in choices of vocabulary gradually de-subjectifying the ubject. Even if the ubject comes to be seen more and more equivalent to a thing and a potential commodity, we should be careful not to assume that its exchange is determined by some sort of magical "market forces." When talking about selling to others, a transfer was actually often referred to as "helping" others. Listen to this bone banker, for example:

> This thing about prices, it's just kind of operational prices when it's kind of public, it's nothing but what it costs to sort of having made sure that the bone is alright and screened and that kind of thing. So in that way you help one another.

Several bone bankers complained that due to the new EU safety regulations, they could no longer "help" the hospitals they used to:

> Before the new law, we did sell to private clinics too. But that stopped.
> *How come?*
> It is because—I think it's too much bother.

Similarly another bone-bank manager thought it used to be easier:

> [I]t's a problem that we are not authorized to distribute bone. Previously, prior to this new rule, you'd be a gentleman and look in the freezer. If there was a slight surplus you would let them have some. But now we're not allowed to, because you have to be authorized.

References to the gentleman attitude and the discourse of "helping" demonstrate the endurance of a public sector ethos in which bone transfers are conceptualized as a matter of shared problem solving. For some of these bone bankers, the gentleman days are gone, but that does not imply that the new system is more businesslike. It is closer to a bureaucracy. Nonetheless, most bone banks try to comply with the rules and continue to "help" each other by "selling" for whatever price they have decided on. They rarely comment on the notion of "bank" and its monetary connotations, except when surgeons distance themselves from the increased bureaucracy associated with the EU Directive, explaining that they cannot spend all their time as "banking executives."

A pragmatic attitude with distance toward monetary issues characterizes also the "consuming" surgeons. One bank located in a top specialized unit, and thus a large consumer of bone, paid very different prices depending on where the bone came from, but said:

> You know what? We really don't care that much. We pay whatever they ask.... Really, we're in a very difficult situation. We need bone. So we don't quibble over whether it's two Kroner more or less.

The expenses related to the new safety standards coming with the transposition of the EU Directive are covered partly through funds granted under a Danish regulation according to which regional authorities must be compensated for assignments imposed on them by the national authorities. Most bone bankers knew these funds were circulating but knew little about whether they would go to the ward or end up somewhere else.[38] A somewhat lengthy quote from a bone banker who was impersonating the bureaucracy surrounding the bank with various voices illustrates the thinking very well:

> We haven't seen any money yet, right? We're just doing loads of work [laughs]—that's hospital life for you. Then it might be, at the end of the day, you're told "Oh, you've spent that much? How come?" "Well, it's because we've got this tissue law added to everything" "Well, really?!" And then it might be that you get an extra appropriation. But guess what— it's just a kind of *monopoly money*. But I know that in several regions they've been very tough and said "we won't take care of the bone banks before we get funds for it." We've more leaned towards saying "Well, we'd better say it'll cost so and so, in working hours and tests and that stuff," right, and then add it to the bill when people order bone. And then you just hope it'll even out [original emphasis].

The point is that irrespective of businesslike languages of transfer ("price," "selling," "buying"), the exchange of bone reflects a range of competing logics expressed in different discourses ("gentleman," "helping," "bureaucracy," "bother," "monopoly money").

Blind Spots Produced by the Moral Paradigm of Market Thinking

Further understanding of the exchange form and its potential for change, that is, its essential plasticity, can be gained from exploration of how bone bankers contemplate adaptation of new grafting technologies. Three Danish university hospitals have looked into the possibilities of expanding their repertoire by beginning to collect bone from deceased people, so-called cadaver donation. Again the reasons should be sought in the materiality of the bone. When relying on femoral heads only, there is a clear limit to the size of the grafts that can be made. If surgeons have access to bigger bones (as well as tendons) and more sophisticated machinery, they can produce more specialized grafts. The cost of this machinery and the working hours going into having a 24-h response unit to receive cadaveric donations is explicitly considered and often seen as a hindrance. One hospital had planned to embark on cadaver donation several years ago and hoped to recover the current expenses by trading grafts, but the regional authorities did not allocate the necessary start-up funds. At another hospital, the chief surgeon anticipated not only having the expenses covered but a potential source of income too:

> If we could run a bone bank that could also make structural grafts, then I think I can say we'd be the only ones having that in Denmark, at least. We might have some interest in that. I don't know if you follow me in that? It would raise the profile of the ward—would even provide a bit of financing if we could—eh—sell it to others. 'Cause it'd be something you would *not* get for free, definitely. [Original emphasis]

When I asked this surgeon to expand on the profitability aspect, the reply was that though his ward might gain financially, the key motivation was the prestige of it. Nevertheless, it shows that the sale of bone grafts can also be thought of as a financial opportunity. Some would think that this surgeon had forgotten the ban on selling bone. However, in the USA and elsewhere, bone grafts generate a sizeable income for many university hospitals and the agencies processing and distributing them; only the surplus is hidden in the so-called processing fees, compensation schemes, etc.—a model which could easily be adopted in the Danish setting. Importantly the inability to determine accurate "recovery costs" is what facilitates this type of profit generation. The moral paradigm of market thinking makes it inappropriate to introduce calculation standards for recovery costs. As in the blood case above, moral reasoning is part of facilitating capitalization.

Curiously, few of the interviewed surgeons had ever felt the need to import grafts. And, perhaps even more remarkable, despite being aware of the free availability of foreign grafts for sale and despite consciously treating the bone they procured as beyond trade, none of the surgeons had contemplated the occasional purchase of a graft as unethical. They did not know how foreign grafts were procured. Once detached from a body, an ubject can gradually lose its subjecthood and become purchasable. This is what some theorists call commoditization, but the moral agency going into the process is misconstrued if we think it as either covering up the process as a veil or as external to it as a somehow independent form of resistance. Moral agency is integral to the exchange form.

(Ex)changing the Metal Hip

Just like the femoral head, its metal equivalent, the prosthesis, has both use and exchange value. It is produced as a commodity and sold freely across national borders, though the selling is structured differently in different health-care systems.[39] Mostly, prostheses are covered by a perforated titanium layer to increase the induction capacity of the patient's own bone.[40] A used prosthesis is not reimplanted into somebody else, but in principle the titanium makes it profitable to recycle the metal.[41] The prosthesis can be disentangled from the body again in case of replacement surgery or upon cremation because the metal does not melt at the 900–1,100°C used for cremating a corpse. You might even say that the material's endurance at high temperatures makes the prosthesis into a problem in need of handling: the crematoria need to do something with the metal pieces lying in the ashes.

In the 1990s, a surgeon and a metal expert from the Netherlands decided to begin collecting used prosthetic devices from crematoria in Europe. Since 2000, they have been working in Denmark too. One surgeon at a university hospital doing many hip revisions (during which the old prosthesis must be disposed of) had heard about this, and he had looked into the opportunities of selling the used prostheses removed in conjunction with replacement surgery. However:

> Nobody bothers to collect that crap and sell it. I've actually once tried to arrange it—we thought that the staff on the surgery ward could have a Christmas party [laughs] or something, if we collected them. But it turned out the price was so low that nobody bothered.

It appears that money generated on collection of used prosthetic devices was expected to circulate somehow outside the normal hospital economy, as pocket money for the staff but not as income for the hospital. None of the hospital wards collected used prostheses. It might be because the price was too low, as suggested above. Another reason could be that an existing waste disposal system had already taken care of them.

Crematoria can get no better pay for used prostheses than hospitals; however, they *do* face the problem of waste disposal, not least because the number of implants has been increasing during the past decades. The material agency of the prosthesis makes it into a problem of a particular type: it can break the urn. In Denmark, it is common to use degradable urns and pursue an ideal of total decomposition "to ensure that nothing reappears," as one crematorium manager explained. Crematoria wish to avoid placing metal in the urn and then having to remove it from the burial plot 20 years later when a new urn goes into the same place. The various crematoria have tried out different solutions of disposal, and one simply dumped the prostheses in a public dump, a "solution" that was publicly criticized. Therefore, the crematoria association had good reasons to establish a certified collection system, and the offer of the Dutch company came in handy.[42] They offered to set up bins for the metal parts and to come by and collect them—besides paying a price for it!

Making Business in a Nonprofit Realm

In Denmark, crematoria are run by parochial church councils on a not-for-profit basis according to the rules laid out by the Ministry of Ecclesiastical Affairs. In 2001, the Ministry issued a circular stating that the gold in the teeth of a corpse and the precious metals in jewelry put into the coffin belonged to the deceased and should go into the urn, while the metal contained in prostheses should be sold to the Dutch company and the money go to the crematoria. Leaving aside this remarkable distinction between metals, I will focus on why 3 (out of 37) Danish crematoria decided *not* to comply with the new rules. In an interview, one of the three crematorium managers explained that the church council had decided that it could not sell something "belonging to the person." Again the distinction between person and commodity operates and defines what can and cannot be done to ubjects. The metal can be sold prior to functioning as a hip; but it cannot as easily reenter trade after having occupied a space in the human body.

I asked the people who had recently received a prosthesis whether they thought about the prosthesis as part of their body. They mostly replied something similar to this woman: "Yes, it helps me, right, or I hope it does in any case." When then considering how she would feel about it being sold upon cremation, she said:

> Well, if it's lying there [in the ashes] they can take it. I don't want them to—I'd really rather not that—when I'm dead, I don't want them to begin cutting me up [to take it out].

Some of those interviewed asked questions about who received the money from the sale of used prostheses, but as long as the crematoria were not driven for profit, then the prospects of increased efficiency together with the environmental aspects of recycling outweighed lingering opposition. Here again the notion of not-for-profit plays a legitimizing role. Crematoria can be compensated for their expenses, but the users of prosthetic devices deem it illegitimate for crematoria to profit from metal recycling (irrespective of whether people were familiar with the law prohibiting for-profit cremation). Hence, just as it was the case with bone, the selling of (bodily) metal must somehow go into a budget without making anybody richer.

How is the price then calculated—first for the cremation and then for the recycled prosthesis when only compensation of expenses is to be covered? Well, very differently in different crematoria. Some church councils cover all expenses with the income from church taxes, while others think that church taxpayers should not pay for people who are not members of the church and they then charge a price for the cremation to have their expenses "compensated." However, as in the case of the price for bone, costs can be calculated in many ways. Some include fuel, working hours, and rent, while others also include the upkeep of the grounds surrounding the crematorium. There are no standards for the cost to be compensated. The price is "basically a political decision made by the church council" as one crematorium manager put it. Also, for those who do charge a price, every cremation is charged the same, even if the same manager remarked that to cremate a skinny person costs as much as 30 times more than an overweight person "who basically comes with his own fuel." It is never given what goes into the calculation of "reasonable compensation" of expenses.

How does the metal recycling then influence the price of a cremation? The crematoria association receives a single amount from the Dutch company for all the metal collected at Danish crematoria. This amount is divided by the number of cremations each crematorium has carried out. Thereby, it is not the specific prosthesis which is compensated or sold; it is the general level of activity in each crematorium.

The amount is viewed by the crematoria managers as disappointingly low, and the income from recycling does not affect prices for a cremation: the extra money goes into the running of the crematorium as an opportunity for a little extra spending instead, or it is donated to "worthy purposes." The Dutch company, however, is free to use its surplus as any other company. Though functioning as "compensation" in a "not-for-profit" economy, this type of money does generate surplus. However, and this is the key point, care is taken to make it appear different from commercial trade, and this care sets the conditions for the exchange system.

The three crematoria that opposed "commercial" recycling collect the metal parts in separate bins, and when full they bury all the prostheses in a designated area on the churchyard. The metal hip is indeed in between subjecthood and the commodity realm: it is too affiliated with the deceased to be sold, but not so closely affiliated that it is deemed necessary to let it follow the other mortal remains into the urn and thereby ruin the principle of total decomposition.

The Productivity of the Undefined ... Heat

I have shown how the exchange of blood, bone, and metal can be totally uncontroversial but also how commercial exchange schemes activate important boundary work as a result of the ambiguous status of these ubjects. As I explored the whereabouts of metal remains, I came across another type of postmortem recycling that can help us to appreciate even better the productivity of an undefined relation to a human subject and how this ambiguity is activated in a particular sense through confrontation with exchange forms that are seen as market-like. Like metal, heat represents a financial value for Danish crematoria. The ovens consume substantial amounts of fuel, and for years, the Danish crematorium association has wanted to sell the generated heat to local power stations. Politicians have repeatedly turned down these requests on what has been called "ethical" grounds. Notice again how ethics can be presented as a safeguard against the "market," much like what happened in the Swedish example. Then, some years ago, this topic was reframed as an environmental issue. Studies documented that another metal used in the human body—the mercury which was for some decades used in tooth fillings—constituted a health risk when released into the air. To extract and contain mercury, the air must be cooled. Cooling consists of capturing heat in water, which is essentially the same process employed to recycle the heat. If not utilizing the heat, the crematoria would need to release it straight into the air through purpose-built constructions. In light of this, the Minister of Ecclesiastical Affairs considered changing his mind and invited the Danish Council of Ethics to write an opinion on the matter. Ethics was to solve the matter for the Ministry and the crematoria, just as ethics was supposed to handle the matter in Sweden when money entered into exchange schemes.

The council's opinion is an interesting document.[43] It clarifies that the reason for having funerals in the first place is to remind people "that it is a human being, a person, who is buried, not a thing"—and to respect the dead body is also to respect the mourners who might think of the body as "not only representing the deceased" but even "identical with this very person." Material continuity is thus identified by the council as a central rationale behind burial rituals. The council supports recycling of heat and provides the following reasons: (1) recycling of heat is not the purpose of the cremation; (2) cremation necessitates extra fuel, and hence, the heat does not stem from bodies only; (3) the environmental reasons for recycling provide evidence that bodies are not perceived of as firewood—heat is only a derivative; and (4) it is a basic condition of "the cycle in Nature" for human beings to "enter into an impersonal circulation, where energy is utilized without the source being viewed as of particular importance." In its conclusion, however, the opinion retracts slightly on its support for recycling. Rather than endorsing free movement within the grand "Cycle of Nature" irrespective of the "source of energy," it is seen as preferable to restrict where this particular type of heat may go. It is suggested that heat from the ovens could be reserved for use in the crematorium and adjacent buildings (i.e., the church room) rather than sold to power stations. This heat, after all, remains different from other types of heat and is seen as better used in secluded spaces outside general circulation and profit-making circles. On closer inspection, the big "Cycle of Nature" appears to consist of several minor cycles that should be kept apart when ubjects are at stake (much like Bohannon's multicentric economy among the Tiv). Finally, the council recommends careful public communication efforts to factually explain how the heat "passes through several separate stages" wherefore there are no leftovers of the deceased in the heat (leaving aside the basic question of the heat itself as representing the person).

As I invited bone donors and recipients to also contemplate the ethics of recycling heat, it was not uncommon for people to either spontaneously laugh or ridicule the politicians dealing with such "absurd issues." Recycling was seen as a "natural" and "rational" activity:

All these immense amounts of heat and they just throw it out—to no use [indignant]! You might as well use it!

or:

If [the heat] can be of any use, I think it's just great. I don't care what it's used for. Really, we might as well—we have all these problems with pollution.

Interestingly, the people finding the problem absurd simultaneously agreed with the ethics council that it might be better to avoid selling the heat commercially. It appears that even heat—intangible or not—has a potential for subjecthood, and again avoidance of commercial trade is an important step toward curtailing this potential. Ethics is seen as a prime resource in the boundary-setting games in this case as it was with bone and metal exchange and in the Swedish handling of blood for genetic research. Ethics policies do not stop exchange practices; they typically facilitate them, as Julia Black has pointed out.[44] But to view the moral agency involved as a mere cover-up for market forces is erroneous. These policies shape what can acquire a price and how, and moral agency is integral to their manufacture and performative effects!

General(izing) Reflections

Market proponents often refer to surveys and other studies ostensibly showing that many people would like it to be legal to sell their body parts and act as, for example, "kidney vendors" in the same manner as I described in Chap. 2 that Ted Slavin sold his blood for research.[45] If they were right, we should have expected also the donors above to be much more eager to make a profit on their blood, bone, and metal parts. Since I want to suggest that the fact that they were not all that eager to sell ubjects has a more general salience outside Scandinavia, I now wish to relate the analysis above to the literature on donor interests that market proponents often use to make their case and extend this discussion to some more general reflections on what donors see as at stake in ubject exchange.

"Vendors Are Themselves Anxious to Sell"

A much cited *Lancet* article from 1998, for example, claims that "vendors are them-selves anxious to sell"[46] and support the claim with a reference to a book chapter in which it says: "Who are the sellers? No one who is comfortably off will undergo surgery, lose time from his work and suffer pain for someone he does not know. It is always the poor who sell their organs."[47] Whether or not this statement delivers support for the "anxious-to-sell" claim is a formal logical topic of some interest in its own right, which I will have to leave for philosophers better trained for that to deal with. The search for evidence of vendor eagerness, however, is a significant element of the debate and demands further attention. Basically, it has to do with what so-called market forces are supposed to do and for whom.

An early study also sometimes referred to in support of the claim that people would like it to be legal to sell their organs was conducted by Astrid and Ronald Guttmann and published in 1993.[48] In this survey (which had a response rate of 47%), 63 members of the general public and 239 medical professionals responded to two cases in which buying a kidney was said to be the only alternative to either death or serious complications from dialysis. When the only alternative was death, 49% of the respondents thought the patients should be allowed to purchase a kidney and in the case of serious complications from dialysis 40% thought so. Irrespective of the poor reliability—it was a small survey with a large drop out—it is interesting that the authors conclude that it indicates great support for a "future market." Not only did the survey present a case in which people were invited to take sides with a person facing death and consider only one alternative solution; its conclusions also seemed to overlook the fact that, according to data published in the same article, three quarters of the respondents thought that the health services and not the recipient should pay (which seems to indicate that access to a kidney should not reflect wealth), that trade boards should not be involved in the regulation, and that transactions should be the responsibility of not-for-profit agencies. To take this survey as indicating public support for a "future market" involves using the plasticity of the concept of market to its full.[49] As I read the survey results, about half of the respondents express some element of willingness to help a dying (wo)man as long as it does not look like a "market" and as long as wealth does not translate directly into health. One might wonder how the results would have been had the respondents also been invited to side with a potential "vendor" in a situation they themselves might one day face, for example, by considering whether potential creditors should be in a position to claim a kidney in case a person cannot pay the mortgage. The example is not all that far-fetched considering how Goyal et al. found that 96% of the 305 Indian real-life "kidney vendors" they interviewed had sold their kidney primarily to pay off debts [50] and how Lawrence Cohen found that money lenders in Chennai in India would consider people's health and donor options before lending them money.[51] Should body parts be considered mere property in a formal legal way in the USA and Europe, then indebted American or European citizens would be facing similar situations.

In 1993 and 2005, the Gallup Organization conducted two surveys on attitudes in the American public toward organ donation which were much larger than the Guttmann study.[52] The 2005 survey followed up on the 1993 survey, and it indicated an interesting development in attitudes to financial incentives. An increasing number (16.5% compared to 12.0% in 1993) stated that they would be more inclined to donate their own organs if paid an incentive. There is a gender aspect to this with more men (21.5%) than women (13.3%) taking a positive view of financial incentives. There is also an interesting polarization in that more people thought financial incentives would make them *less* likely to donate (from 5% in 1993 to 8.9% in 2005). Though market proponents often compare tissue and organs to personal property, a similar study on people's right to sell land, labor, or consumer goods would never bring about such results—less than 17% *considering* a monetary incentive to do some-thing. We do not know whether they would really go ahead with it and even more importantly whether they would consider it a preferred option. Nor do we know how people would feel about a creditor being in a position to demand sale of an organ. I think few people would like the risk of debt to reach below their skin. In fact, I would warn strongly against reading such surveys as indicative of 16.5% of the American population thinking that it would be just swell to sell a kidney if only they were allowed. The rising number of people thinking that they might have to consider such an option nevertheless indicates a gradual change in norms. Studies conducted in Europe indicate a more uniform opposition to financial incentives, and in one study the main point for potential donors was that demand should not control supply; rather the role for policymakers should be to ensure fair distribution of available organs, not to make people donate who did not want to do so.[53] Despite a general policy inclination to go against any form a remuneration of donors, there are some European examples of changing attitudes in policymaking circles, in particular with respect to monetary incentives for organ donation.[54] The moral paradigm keeping the body and its associated ubjects out of commercial exchange relations is a historical construct, and it will at some point evolve into something else. The most pressing question in the short term is what role we want policymakers to play in the process.

What Might Be at Stake for Most People

Some readers might think that the way in which I have described donor attitudes in the two cases above must be a very Scandinavian phenomenon irrelevant for understanding what is at stake in, at least, the USA. Some might even think of Ted Slavin or John Moore as prominent examples of Americans trying to profit from the selling of what they themselves construe as "their body parts" (see Chap. 2). Certainly there is a lot of ethnographic specificity to the two examples of ubject exchange outlined above, and the survey data further solidifies this point. However, we should also be careful not to put too much emphasis on cultural difference and assume that a propensity to "sell" or to refrain from selling is just a matter of cultural context, as if Americans and Scandinavians live in totally different worlds. To get behind the differences seen in the survey data, we need to contemplate what is at stake for the people engaged in ubject donation and how these interests relate to the institutionalized ability to ignore monetary elements of the donation. Concretely, I wish to suggest that many potential donors will assess compensation schemes in light of *available alternatives* for income and the *relative sense of control* they feel (and not just their "culture").

John Moore and Ted Slavin might be seen as people opting for ubject exchange as their preferred source of income.[55] Ted Slavin did trade his cells to some companies; however, as also pointed out in Chap. 2, he also gave the cells for free to the one researcher he really believed would make a difference in medicine. He was not all that different from the Swedish blood donors. John Moore, on the other hand, felt unfairly treated and really went for the cash, not the public health contribution. His monetary ambitions, however, were sparked by an act of deceit. There is a strong cultural tendency in the USA for translating wronging into monetary compensation, so much that accidents are occasionally talked about as opportunities for economic gain.[56] The tort system institutionalizes this particular approach to wronging, but it is not a uniquely American approach (in fact, ancient legal systems often produced lists of compensation for losses of body parts such as Codex Hammurabi in Babylonia and Æthelberht's early British code). What Moore's case says about the difference between the USA and Scandinavia relates more to differences in the legal culture of tort, I contend, than to differences in how eager people in general are to see ubject exchange as a primary source of income (besides, Moore's attempt to undo wronging through monetary compensation is not necessarily a reason for actively facilitating monetary *incentives* for ubject donation).

There are also other differences between the USA and Scandinavia in terms of monetary incentives for ubject donation, and I will expand upon these in the following chapter. While most compensation schemes of donors have been brought to a halt in Sweden and Denmark, many Americans are invited to consider their bodies as a potential source of income for periods of time as they are enrolled as egg or sperm donors, or deliver breast milk or blood plasma. Nevertheless, many of my Swedish informants had received money once and not felt offended. While market proponents sometimes use this to argue that all forms of ubject trade are a donor interest,

I believe ubjects can change hands for money without causing harm only as long as donors do not feel their phenomenological sense of control challenged. When some claim to find their own experiences with trade utterly unproblematic, I contend that they have retained a sense of control. Of course, one might have a feeling of control without actually controlling the risks incurred.

Like John Moore and Ted Slavin, we saw how a few Swedish respondents said that they thought they should be free to ask for money for their blood. Importantly, however, they would prefer giving it for free if they believed in it bringing genuine medical progress (again similar to Ted Slavin). Since they did not find this type of progress a likely outcome of the proposed research, they did not donate at all, and they would not have liked to be a in a position where they *had* to sell their blood to get the money. Why did they insist on a right to sell their blood? I believe that they used the reference to property rights to communicate a right to retain control. The interviewed Danish bone donors were against installing incentives that could put their health at risk (a surgeon snatching a little too much bone), but they did not mind the hospital getting its expenses covered. Again, they focused on control, rather than monetary concerns as such.

Anywhere, when people get poor or desperate enough, taking risks with one's body to gain material resources can appear the least unappealing of available options. Often enough other people see an opportunity in their suffering, as Veena Das remarks;[57] often enough there are people willing to pay for the risks taken by others.[58] Money is a form of power—it is a medium for negotiation between conflicting wills—and if it can buy access to bodies, somebody will want to use their monetary power for that purpose. All the same, when Nancy Scheper-Hughes encounters poor shanty town dwellers in Brazil who are eager to sell a kidney, we would be fooling ourselves if assuming it is their preferred way of income.[59] When market proponents make claims about vendor eagerness, they are in most cases illustrating the effects of dedicated not knowing about the living conditions that make, for example, kidney selling appear as the best way out of a crisis.

Theft and the Affordances of Not Knowing

The sense of control is related to trust. Moore experienced a breach of trust and opted for property rights. Trust is also at stake in a different set of stories through which we sometimes learn in the media about the everyday practices of ubject exchange, namely, theft. Stories of theft are important in bringing back to light what used to be hidden, and they raise all the uncanny feelings characteristic of ubject exchange. In a widely broadcasted scandal from Nuremberg (of all places), three German crematoria workers collected gold—primarily from teeth—and made approximately 150,000 Euros on it. The nature of the wrongdoing was difficult to ascertain, however. The German crematoria workers were first acquitted because the gold represented part of persons' bodies, and as such it could not be owned, and when not owned, it could not be stolen. In an appeal case, however, they were convicted for desecration of graves and for breach of obligations toward their employer.[60] Similar cases are reported from the UK, though not on the same scale.[61] Once out in the open, such narratives of theft disturb the peace otherwise provided by that we do not know. They also interfere with the sense of control. We might say the "sellers" are convicted for bringing back to light the existence of "buyers," and yet such cases rarely seem to question the legitimacy of buying an ubject as such. All the same, when gold can be sold, some will occasionally grab the gold at their disposition.

The same goes for biological ubjects. Irrespective of the material composition of the ubject, not knowing where an ubject comes from facilitates its travel. Think of, for example, the fertility doctor Ricardo Asch who harvested eggs without consent from one patient and provided them to another who posed no questions.[62] Or think of the scandals surrounding the UCLA willed body program in which Director Henry Reid was making a profit on selling donated body parts to research companies[63] or of the American company Donor Referral Services which was caught faking death certificates in a case similar to that surrounding Alistair Cooke as outlined in Chap. 1[64] or of the publicly employed British surgeon Christopher Ibbotson, who snatched bone from his daytime job and sold it to a processing facility.[65] Such cases share elements with the widely reported illegal transfer of organs, and yet they differ by bringing the drama to known facilities used by middle-class Westerners. When, however, such stories are used to illustrate that "there *is* a market, and to say that there is not is to perpetuate a fiction," there is reason to pause. We are dealing with stories of illegal acquisition, not "markets" in the sense suggested by market thinking. These recurrent narratives do, of course, indicate a strong interest among some people in selling ubjects—but we should remember they are not selling ubjects originating in their own bodies. Narratives of theft are therefore not documenting "market forces" that can be released so that the "invisible hand of the market" will unproblematically supply the ubjects desired by contemporary biomedicine. They do not document a huge "body reserve" eager to fulfill demand once prohibitions have been lifted. They illustrate that if it is legitimate to buy a given ubject, some-body will probably at some point find a way to procure it, legally or illegally. Buying makes theft an option.

Revelations of theft thereby inform us about this one particular and yet quite important aspect of the normal, mundane everyday practice of ubject exchange. People "buying" ubjects do not always feel obliged to know about their (potentially illegal) origin. Stories of theft illustrate how ubjects can be acquired for money without raising suspicion. Once the ubject is properly disentangled from donors, it appears that exchange partners find its monetary valuation increasingly easy and, sometimes, appealing. For example, it seems unproblematic for a surgeon to imagine *buying* the bone needed to treat a patient. The fact that the same people who view the ubjects they harvest as clearly beyond trade can concurrently consider purchasing other ubjects they do not know the origin of is very telling. It explains the preconditions for theft. A will *not to know* produces a system where the awkward few can grab bone, tendons, and precious metals they were not supposed to possess and sell them to "customers" who feel little urge to explore the origins of the desired ubject.[66] Not knowing facilitates a (unfounded, some would say) sense of moral integrity.

Market proponents typically say that full-blown legalization of ubject trade will bring an end to black markets.[67] It is most unlikely. Just as "black markets" did not disappear in Eastern Europe despite the cheerful prophecies of market proponents prior to the fall of the socialist regimes,[68] legalization of monetary compensation in this field will not erase illegal ubject transfer. The very ability to sell ubjects without being asked unpleasant questions will continue to tempt surgeons, undertakers, and others having access to unprotected bodies to take that which was not intended for trade. The practice of "buying" builds on a strong tradition for (legitimate) .not knowing the origin of the ubject. This mechanism does not disappear just because ubject trade becomes legally endorsed, and therefore illegal acquisitions will not disappear as a consequence of increased legitimacy attributed to the act of oblivious "buying." Market proponents are partly right, however, when they accuse the fear of commoditization for creating a systematic blindness with respect to monetary issues, and market opponents would do well by listening to the critique.

Lessons Learned from Everyday Practices

With this chapter, I have illustrated how blood, bone, metal, and heat all acquire a potential for subjecthood through association with the human body and shown how this potential is most fiercely activated when the ubjects are enrolled in exchange forms that are seen by the involved actors as somehow market-like. As ubjects, they embody a fruitful darkness seething with partial connections. I have included ubjects which challenge prevalent assumptions in the literature about what constitutes a "human body part" to show that also metal and intangibles can acquire traits mostly discussed in relation to "tissue economies." The point is that we should not see these traits as residing in "tissue" as an ontological category but in their relationship to experiences of subjecthood and worthiness. Also, I have suggested that moral reasoning informed by the moral paradigm of market thinking—which seeks to purify the distinction between persons and commodities—is operative in creating elements of *not knowing* and that ethics policies embody this type of moral agency and participate in the production of not knowing by fixating the ethical gaze on manageable problems. What is seen as manageable differs depending on the reasons for engaging in exchange. When blood is desired for its informational properties (a DNA code), blood as a substance is bifurcated from blood as information. When bone is desired for its implant properties, the hip is bifurcated from its processing. In both cases, the Lockean notions of commercial benefit as a reward for labor (as discussed in Chap. 2) are operative, and the donors from whom the ubjects flow are construed as passive holders of material and beyond the realm of economic compensation. Rather than writing off the ingenuous agency going into the establishment of these strange modes of exchange as merely a "denial" of the "truth," as some commoditization theorists seem to do, we need to take this agency seriously for what it does. I do not think all this work is just a matter of depriving donors of money. It is about guarding important societal boundaries.

Moral reasoning, in particular in the form it takes in ethics policies, is not just about protecting or safeguarding; it is part of shaping exchange systems and designing regimes for accumulation. Who may benefit in which ways? Who are entitled to what? In as far as ubjects cannot be "owned," new types of entitlement must be created. Ubject exchange thereby potentially transfigures one of the legitimizing components of market thinking. In place of ownership, new forms of entitlements are created: informed consent becomes one type of entitlement (vested in the donor) which is distinguished from the entitlement to dispose of the donated ubject (vested in the biobank), the entitlement to dispose of samples (vested in the biobank) becomes distinguished from the entitlement to negotiate intellectual property rights (vested in the company), and entitlements to collect bone (vested in the public hospital) are distinguished from the right to profit from processing it (vested in the private company). Market thinking is not immune to the hybridization which the current biomedical advances involve, and, therefore, capitalist exchange and its property structures are multiplying and transforming in the process of handling an increasing number of ubjects. Furthermore, it must be appreciated that exchange goes beyond mere movement of stable entities. In the fruitful darkness of ambiguity many changes take place. Ubjects lose or gain properties thanks to the exchange schemes in which they move. For example, they become objectified or subjectified just as their physical properties change as they are incorporated into databases or processed into implants. Exchange implies change.

This chapter has shown how there are some *similarities* across various types of ubject exchange in terms of obscuration of the gradual price-setting process. Great care is taken in making the transition from part of subject to commercially available object go unnoticed. The chapter also illustrated clear *differences* in terms of the amount of attention various ubjects get from the public and from policymakers. For policymaking and scholarly thinking to take account of such differences and to avoid building universal policies based on specific ubject types or situations, we need to reflect more basically on what it is that makes some material flows controversial and others so mundane and almost too ordinary to make anybody bother. This is the topic of the following chapter.

Endnotes

1. Price setting in fields other than ubject exchange illustrates mechanisms similar to what I am about to show at play. Price setting of tribal art, for example, needs to balance connection to the location of origin with effective relocation (it must not be too local to be sold and yet local enough to carry ethnographic value), and price setting in fair trade faces ambiguities similar to the calculation of compensation and recovery costs in ubject procurement agencies (in terms of attempt of reaching a "fair" price without relying on the supply-demand dictum of market thinking). See Geismar (2001), Varul (2008), and Chapter 2. However, my point is not to build a theory of price setting, nor to claim a special case pertaining to all ubject exchange. I merely mean to show how

an ubject's unruly potential can have special effects when certain societal institutions are at stake.

2. For example, Anderlik and Rothstein (2001), Beauchamp and Childress (2001), Cambon-Thomsen et al. (2007), Gillon (2003), Hansson (2005), and Knoppers et al. (2006).
3. Amit (2000), Bosk and De Vries (2004), Brekke and Sirnes (2006), Cooter (2000), Corrigan (2002, 2003), Das (1999), de Vries et al. (2006), Epstein (2009), Hayden (2007), Hedgecoe (2004), Kelly (2003), Lundin (2002), Pálsson and Rabinow (2005), Scheper-Hughes (2001a), Scocozza (1994), and Whitt (1999).
4. I have reviewed the anthropological contribution to ethics elsewhere; see Hoeyer (2006). Other attempts of finding a middle ground can be found in, for example, López (2004), Zussman (2000), and Metzler (2010).
5. Elliott (1999).
6. Last (1981:387). Similar points were made by Bateson when arguing that noncommunication was constitutive for meaning production (Bateson 1972).
7. In her survey of the tissue procurement field, Michel Goodwin found more than 200 for-profit companies handling tissue and organs in the USA. She provides the example of a company paying the Coroner's Office $250 for a set of corneas while taking $3,400 when delivering them after "processing" to a transplant unit (Goodwin 2006:17).
8. The academic debate about deCODE is quite extensive. Good places to start are found in Árnason and Simpson (2003), Fortun (2008), Pálsson (2002a, b), Pálsson and Harðardóttir (2002), and Rose (2001).
9. By 2010, 140,000 Icelanders (out of a total population of approximately 320,000) had donated blood to the biobank, which also contains blood samples from major research projects conducted elsewhere and therefore has access to samples from additional 350,000 people (Moldestad 2010).
10. See discussion in Arnason (2004), Merz et al. (2004), and Potts (2002).
11. My fieldwork took place primarily in and around the town, Umeå and on two locations in the interior region. As an intense conflict over entitlements to the biobank evolved between different stakeholders, the conflict began featuring in local and national newspapers. Subsequently, UmanGenomics has failed in attracting investors, and, during the spring of 2003, the employees of UmanGenomics were laid off. I followed the development of the conflict through contacts with both employees of UmanGenomics and the Medical Biobank, and throughout the study asked my informants who they saw as key actors in the conflict and in relation to the issues of the ethics policy, and pursued interviews with these persons. The interviews with donors and nurses were conducted during the autumn of 2000, the spring of 2002, and the spring of 2003, tape-recorded and transcribed. I am responsible for all translations. With colleagues, I also conducted two surveys which have been published separately (Hoeyer et al. 2004, 2005).
12. Abott (1999) and Nilsson and Rose (1999).
13. Scheper-Hughes (2001a:2).
14. For a discussion of these elements of Swedish ethnography, see Frykman and Löfgren (1987), Nordberg (1998), and Qvarsell (1986).
15. See, for example, the framing by medical professionals in Gullbring (1952) and the official statements in Kungliga Medicinalstyrelsen (1956).
16. Socialdepartementet (1995:831).
17. Rabinow (1999:96).
18. Socialdepartementet (2002:297).
19. Analysis and more details about the conflict can be found in Hoeyer (2004) and Rose (2003).
20. See Gold (1996) for an interesting analysis of this history. Anthropological analysis of ownership forms of relevance to an understanding of the specificity of the gene patent includes Hann (1998), Pottage (2004), and Strathern (1999). Elsewhere, based on a reading of patent court cases, I have suggested that the moral paradigm of market thinking can be seen as manifesting itself in patent history as a series of attempts of keeping the body as a substance separate from information as a result of work (Hoeyer 2007).
21. Hoeyer (2002), see also Andreasen and Hoeyer (2009) and Einsiedel and Smith (2005).

22. It has been discussed in detail elsewhere how informatization of biological knowledge facilitates accumulation, capitalization, and propertization; see, for example, Parry (2004a, b), and Thacker (2005).
23. This point about separation of two exchange systems has also been made by Warwick Anderson in his analysis of exchange of brains for research (Anderson 2000); see also Mitchell and Waldby (2010) and Tutton (2004).
24. Franklin (2003:123).
25. Merx et al. (2003).
26. Tomford (2007) and Wilson (1947). The earliest report of a successful bone transplant is probably 1668 (McNamara 2010), but today's bone transplant research took on following Macewen's publication of his 1878–1880 experiments on a 10-year-old boy (Macewen 1881).
27. For UK figures, see Advisory Committee on the Safety of Blood, Tissues, and Organs (2007), and for USA figures, see Vangsness et al. (2006).
28. Anderson and Schapiro (2004), Cheney (2006), Holtzclaw et al. (2008), and Joyce (2005).
29. In the USA, new standards have been introduced in the Food and Drug Administration (FDA) regulation and Section 361 of *The Public Health Service Act* (Joyce 2005), while the *EU Tissues and Cells Directive* restructures bone banking and other therapeutic tissue collection practices in the European Union (Hoeyer 2010).
30. My fieldwork in Denmark was conducted intermittently between February 2007 and January 2009. Inspired by Marcus' (1995) multi-sited ethnography, I followed objects moving in and out of bodies across various sites and interviewed regulators in European and Danish agencies, commercial bone processors, Danish bone bank managers, surgeons, donors, and recipients. After I had undertaken initial pilot interviews, my assistant Sofie Okkels Birk interviewed all 17 Danish bone bank managers (ten of them twice). My participant observation focused on informed consent procedures in relation to donation of bone at a Danish hospital on seven nonconsecutive days and participation in meetings among bone bankers. I in-depth interviewed seven donors and four recipients of bone. I also taught at courses for bone bankers and used the feedback during these courses as data. Concomitantly with my study of bone banking, I visited crematoria—observing the disposal of metal remains—and followed the policy process and public debate concerning recycling from crematoria. I interviewed people working in and with crematoria management as well as the commercial actors responsible for the recycling. I translated Danish quotes into English at the stage of writing (rather than prior to the analysis) and sought to find appropriate, corresponding colloquial expressions. I have enjoyed the help of professional translator Carol Bang-Christiansen in this, but all inaccuracies remain my responsibility.
31. See also Fontein's discussion of the ambivalent agency of bone in Zimbabwe where parameters of agency are very different from the Danish medical setting (Fontein 2010).
32. Callon (1998:19).
33. Approximately 80% of the hip replacements are done on indication of osteoarthritis, which is an illness of the cartilage, not the bone, wherefore the bone can be used in other patients.
34. Hetherington (2004) and Thompson (1979).
35. There is a lot of similarity between the work described here and the work described by Lynn Morgan in her analysis of the history of embryo collections in the USA, in particular in the ways in which the people collecting ubjects need to be taught to reevaluate ubjects from waste to value when put into the right (medical) hands (Morgan 2009).
36. Since 2001, elective surgery is to a great extent moved into private hospitals compensated by the regional authorities financed by taxes.
37. In fact, quality and weight considerations have gone into the bone-collecting practices in some units but without affecting the price. At two hospitals, for example, knees used to be collected too, but it was found that considering the limited weight and size of the bone, the cost of testing was too great. At another hospital, it was decided not to collect femoral heads from donors weighing less than 45 kg as their bones would be too small to warrant the effort of testing. However, such considerations are not made in the idiom of market value or profitability.

38. Nobody wanted to give me the actual report containing the information about how the funds were calculated until one informant passed me the report number in the official filing system (saying that I should not say where I had it from). I could then use that number to request a copy of the report from the official registry. It was on public record and should be available to the public (only you would need to know its exact number for the registry to locate it), but everybody seemed to feel that it contained knowledge it was better not knowing about. The report itself contains an interesting range of mistakes and uncertainties concerning bone transplants, but the fact that it seemed important for so many people to retain a sense of not knowing about how the compensation was calculated was more interesting for me than the not knowing manifested in the report itself (see Indenrigs-og Sundhedsministeriet 2006).

39. In the USA, there is direct-to-consumer marketing and some amount of brand awareness, while in Denmark, the prosthesis is chosen by the surgeon; see also Anderson et al. (2007).

40. Agrawal (1998).

41. Pacemakers, conversely, are recycled. I know of one group of Danish surgeons from the city of Viborg bringing second-hand (or second heart, perhaps) pacemakers to Bolivia where they operate on the local residents (thank you to Dr Louise Engell for informing me about this practice).

42. There is a long tradition for studying how the solutions at hand contributes to the framing of the problem addressed; see, for example, Koch and Svendsen (2005) and Spector and Kitsuse (2001) or neo-institutional theory (Lindblom 1959; March and Olsen 1976; Meyer and Rowan 1991).

43. Det Etiske Råd (2006). The council was surprised to be presented with this assignment which is outside its formal jurisdiction (healthcare and research) and initially even found the task somewhat absurd (p.c.).

44. Black (1998).

45. Presentation of survey data indicates an interest in having the "public" on one's side. A review of studies of public attitudes to compensation for tissue donation is provided by Sally Satel (2008). It is interesting reading as many studies are read by Satel as supporting in particular kidney sale, even though it takes a somewhat vested reading to reach this conclusion (I cannot go through every study and Satel's interpretation here). Even if Satel's reading is overly optimistic in terms of the public support she hopes to find for her own preferred solution to kidney shortage, it is clear that the public is not uniformly against compensating living kidney donors.

46. Radcliffe-Richards et al. (1998:1950).

47. Mani (1992:165).

48. Guttmann and Guttmann (1993). Even a very reasoned philosopher as Stephen Wilkinson uses this article in support of a "market model" (Wilkinson 2003), though sometimes one wonders if the scholars citing it have really read it.

49. Similar vested readings of one's own survey results are characteristic of much of the subsequent work on this topic (Bosisio et al. 2011; Boulware et al. 2006; Bryce et al. 2005; Cantarovich et al. 2007; Cosse et al. 1997; Kranenburg et al. 2008). Interestingly, such studies are practically always conducted by people affiliated with institutions representing the interests of organ recipients. Only one study has found strong support for monetary compensation, and it was conducted in the Philippines (Danguilan et al. 2012). Even here, where 96–98% would appreciate something in return for donated organs (e.g., tax credits or reimbursement of funeral expenses), only 31% are in favor of cash payments. Furthermore, a majority (63%) wanted reimbursement to be called a "token of gratitude." It appears they want exchanges to take a form quite unlike that prescribed by market thinking.

50. Goyal et al. (2002). The average family income fell by a third following the nephrectomy due to subsequent health problems, and 79% would not recommend others in a similar situation to do the same.

51. Cohen (1999) and Cohen (2005).

52. The Gallup Organization (2005). With telephone interviews, 2,500 people older than 18 years were included in the survey.

53. Schicktanz and Schweda (2009). Studies looking into public attitudes in Europe to monetary incentives for tissue donation for research indicate that people prefer doctors giving them relevant feedback and that they would not want money (Felt et al. 2009; Haddow et al. 2007). A study from Canada also indicated that people wanted feedback, not money, for their participation (Godard et al. 2007). Following my argument above, selling their tissue would be seen as disengaging the researchers from obligations. See also discussion in Bister (2010).

54. During the spring of 2010, the Nuffield Council held a public consultation on attitudes to remuneration of organ and tissue donations (Nuffield Council on Bioethics 2010), but the result was to continue a procurement strategy that did not rest on monetary incentives. In the Netherlands, various policy initiatives have contemplated remuneration, following a much debated television hoax called the Big Donor Show in which an actress pretended to have a brain tumor and was made to choose between three potential recipients of her organs upon her death. In the UK, authorities have also suggested removing restrictions of remuneration of gamete donors (Palmer 2010).

55. In several cases, anonymous reviewers from American journals have suggested that my Swedish case material had little to offer an American context using these two examples to suggest that Americans more broadly would be eager to sell their ubjects. Their almost iconic status in the literature makes them apt for the discussion I now begin.

56. See discussion in Oakley (2007).

57. Das (2000). Though Das makes her case based on Indian material, it is fair to say that it is also the case in wealthy nations such as the USA. Think of, for example, the case of alleged organ sale through MatchingDonors.com discussed by Steinbrook, where the supposed vendor acquired 5,000 USD as a "compensation" for an unrelated donation. His "altruistic" donation just happened to coincide with his need to raise this money to pay child support in case he should not lose the right to see his kids (Steinbrook 2005). In low-income countries, planning is a privilege granted only the few, and here, taking any "opportunity" coming along is part of the daily struggle to stay alive (Biehl 2011; Johnson-Hanks 2005).

58. Individuals will desire bodies for sex, violence, or other types of "pleasure," just as organizations will continue to desire bodies as research tools, data sources, or as workforce. The preferable option is typically a body that does not complain in case of incurred harm.

59. Scheper-Hughes (2000, 2001b, 2005).

60. Justiz in Bayern (2008), see also Bild (2006), Kanal 8 (2009), and Radio Charivari (2007).

61. BBC News (2009).

62. Kellerher and Christensen (2005).

63. Ling (2010).

64. Brenner (2006) and The Food and Drug Administration (2009).

65. BBC News (2005) and Govan (2005), the Crown Court of Sheffield ruled Ibbotson guilty of obtaining property by deception on January 19, 2005 (case no. T20040394). He was sentenced to 10 months of prison.

66. Due to the nature of the illegal act, it is difficult to assess the amount of "thefts," but thefts have been reported in medicine from grave robbers supplying dissection schools in eighteenth-century Britain to body snatching from, in particular, black cemeteries well into the early twentieth century in the USA (Goodwin 2006; Moore 2005). Dr Jekyll and Mr Hyde coexist in the benign pursuits of science and the willingness to pay for unprotected cadavers.

67. Becker (2009) and Goodwin (2006). One alternative form of this assertion is that black markets would disappear simply because the people who could not acquire an organ legally would be operating without insurance or public coverage, and as they would have to pay a higher price to acquire it illegally, they would not be in a position to afford it (Taylor 2008). This position with its focus on allocation ignores that organs could be *procured* illegally, probably cheaper, if simply taken without paying donors. And it ignores how organ transplantation will still have to involve an element of medical allocation/triage that some patients would like to circumvent. Elsewhere Taylor discusses procurement through organ theft and asserts that there is no reason to believe any of the stories about illegitimate acquisition described in the literature (Taylor 2005:197). It is of course a

lovely illusion, but it is too naïve to be considered seriously. It illustrates instead how dedication to a so-called market model can determine what can be acknowledged as facts.
68. Dunn (2005). See also the humorous discussion in Amann (2003).

References

Abott A (1999) Sweden sets ethical standards for use of genetic "biobanks". Nature 400:3
Advisory Committee on the Safety of Blood, Tissues and Organs (2007) Overview: delivery and regulation of UK transfusion and transplant services. Department of Health, London
Agrawal CM (1998) Reconstructing the human body using biomaterials. J Mater 50:31–35
Amann R (2003) A sovietological view of modern Britain. Polit Q 74(4):468–480
Amit V (2000) The university as panopticon: moral claims and attacks on academic freedom. In: Strathern M (ed) Audit cultures: anthropological studies in accountability, ethics and the academy. Routledge, London, pp 215–235
Anderlik MR, Rothstein MA (2001) Privacy and confidentiality of genetic information: what rules for the new science? Annu Rev Genomics Hum Genet 2:401–433
Anderson W (2000) The possession of kuru: medical science and biocolonial exchange. Comp Stud Sci Hist 42(4):713–744
Anderson MW, Schapiro R (2004) From donor to recipient: the pathway and business of donated tissues. In: Youngner SJ, Anderson MW, Schapiro R (eds) Transplanting human tissue: ethics, policy and practice. Oxford University Press, Oxford, pp 3–13
Anderson J, Neary F, Pickstone JV (2007) Surgeons, manufacturers and patients: a transatlantic history of total hip replacement. Palgrave Macmillan, New York
Andreasen M, Hoeyer K (2009) DNA patents and the invisible citizen: the role of the general public in life science governance. SCRIPTed 6(3):538–557
Arnason V (2004) Coding and consent: moral challenges of the database project in Iceland. Bioethics 18(1):27–49
Árnason A, Simpson B (2003) Refractions through culture: the new genomics in Iceland. Ethnos 68(4):533–553
Bateson G (1972) Steps to ecology of mind. University of Chicago Pres, Chicago
BBC News (2005) Health manager sold human bones. BBC News (online)
BBC News (2009) Precious metal "sold" after cremation. BBC News (online)
Beauchamp TL, Childress J (2001) Principles of biomedical ethics. Oxford University Press, Oxford
Becker G (2009) Allowing sale of organs will increase the number of donations. In: Egendorf LK (ed) Organ donation: opposing viewpoints. Gale Cengage Learning, Detroit, pp 61–67
Biehl J (2011) Homo economicus and life markets. Med Anthropol Q 25:278–284
Bild (2006) Zahngold-Raub im Krematorium, 23 Oct 2006 (online)
Bister MD (2010) Soziale praktiken des einwilligens: informed consent-verfahren und biomed-izinische forschung im krankenhauskontext. Universität Wien, Wien
Black J (1998) Regulation as facilitation: negotiating the genetic revolution. Mod Law Rev 61(5):621–660
Bosisio F, Santiago M, Benaroyo L (2011) Financial incentives to improve organ donation: what is the opinion of the Vaud French-speaking population? Swiss Med Wkly 141:w13312
Bosk CL, De Vries RG (2004) Bureaucracies of mass deception: institutional review boards and the ethics of ethnographic research. Ann Am Acad Polit Soc Sci 595(1):249–263
Boulware LE, Troll MU, Wang NY, Powe NR (2006) Public attitudes toward incentives for organ donation: a national study of different racial/ethnic and income groups. Am J Transplant 6(11):2774–2785

Brekke OA, Sirnes T (2006) Population biobanks: the ethical gravity of informed consent. BioSocieties 1:385–398

Brenner G (2006) FDA investigates human organ business. ABC News, 21 Aug 2006 (online)

Bryce CL, Siminoff LA, Ubel PA, Nathan H, Caplan A, Arnold RM (2005) Do incentives matter? Providing to benefits to families of organ donors. Am J Transplant 5:2999–3008

Callon M (1998) The embeddedness of economic markets in economics. The laws of the markets. Blackwell Publishers, Oxford, pp 1–57

Cambon-Thomsen A, Rial-Sebbag E, Knoppers BM (2007) Trends in ethical and legal frameworks for the use of human biobanks. Eur Respir J 30:373–382

Cantarovich F, Heguilén R, Filho M, Duro-Garcia V, Fitzgerald R, Mayrhofer-Reinhartshuber D, Lavitrano M, Esnault V (2007) An international opinion poll of well-educated people regarding awareness and feelings about organ donation for transplantation. Transpl Int 20:512–518

Cheney A (2006) Body brokers: inside America's underground trade in human remains. Broadway Books, New York

Cohen L (1999) Where it hurts: Indian material for an ethics of organ transplantation. Dædalus 128(4):135–166

Cohen L (2005) Operability, bioavailability, and exception. In: Ong A, Collier S (eds) Global assemblages. Blackwell, Malden, p 79

Cooter R (2000) The ethical body. In: Cooter R, Pickstone J (eds) Medicine in the twentieth century. Harwood Academic Publishers, Amsterdam, pp 451–468

Corrigan O (2002) Trial and error: a sociology of bioethics and clinical drug trials. University College London, London

Corrigan O (2003) Empty ethics: the problem with informed consent. Sociol Health Illn 25(3):768–792

Cosse TJ, Weisenberger TM, Taylor GJ (1997) Public feelings about financial incentives for donation and concern about incurring expenses due to donation in one US city. Transplant Proc 29(8):3263

Danguilan RA, De Belen-Uriarte R, Jorge SL, Lesaca MRJ, Amarillo MLL, Ampil RS, Ona ET (2012) National survey of Filipinos on acceptance of incentivized organ donation. Transplant Proc 44:839–842

Das V (1999) Public good, ethics, and everyday life: beyond the boundaries of bioethics. Dædalus 128(4):99–134

Das V (2000) The practice of organ transplants: networks, documents, translations. In: Lock M, Young A, Cambrosio A (eds) Living and working with the new medical technologies: intersections of inquiry. Cambridge University Press, Cambridge, pp 263–287

de Vries R, Turner L, Orfali K, Bosk C (2006) Social science and bioethics: the way forward. Sociol Health Illn 28(6):665–677

Det Etiske Råd (2006) Det Etiske Råds svar på kirkeministerens henvendelse om varmegenvinding fra krematorier. Høringssvar, 6 Nov 2006. Copenhagen

Dunn EC (2005) Standards and person-making in East Central Europe. In: Ong A, Collier SJ (eds) Global assemblages: technology, politics, and ethics as anthropological problems. Blackwell Publishing, Oxford, pp 173–193

Einsiedel EF, Smith JA (2005) Canadian views on patenting biotechnology. Canadian Biotechnology Advisory Committee, Calgary

Elliott C (1999) A general antitheory of bioethics. Bioethics, culture and identity: a philosophical disease. Routledge, New York, pp 141–164

Epstein M (2009) Sociological and ethical issues in transplant commercialism. Curr Opin Organ Transplant 14(2):134–139

Felt U, Bister MD, Strassing M, Wagner U (2009) Refusing the information paradigm: informed consent, medical research, and patient participation. Health 13(1):87–106

Fontein J (2010) Between tortured bodies and resurfacing bones: the politics of the dead in Zimbabwe. J Mater Cult 15(4):423–448

Food T, Administration D (2009) FDA public health notification: donor referral services. The Food and Drug Administration, Silver Spring

Fortun M (2008) Promising genomics. University of California Press, Berkeley

Franklin S (2003) Ethical biocapital: new strategies of cell culture. In: Franklin S, Lock M (eds) Remaking life and death: toward and anthropology of the biosciences. School of American Research Press/James Currey, Santa Fe, pp 97–127

Frykman J, Löfgren O (1987) Culture builders—a historical anthropology of middle-class life. Rutgers University Press, New Brunswick/London

Geismar H (2001) What's in a price? An ethnography of tribal art at auction. J Mater Cult 6(1):25–47

Gillon R (2003) Ethics needs principles—four can encompass the rest—and respect for autonomy should be "first among equals". J Med Ethics 29:307–312

Godard B, Marshall J, Laberge C (2007) Community engagement in genetic research: results of the first public consultation for the Quebec CARTaGENE project. Community Genet 10:147–158

Gold ER (1996) Body parts: property rights and the ownership of human biological materials. Georgetwon University Press, Washington, DC

Goodwin M (2006) Black markets: the supply and demand of body parts. Cambridge University Press, New York

Govan F (2005) Lab boss accused of bone thefts. Telegraph, 11 Jan 2005 (online)

Goyal M, Mehta RL, Schniederman LJ, Sehgal AR (2002) Economic and health consequences of selling a kidney in India. J Am Med Assoc 288:1589–1593

Gullbring B (1952) Rekrytering av Blodgivare [Recruitment of blood donors]. Svenska Läkertidningen 49:42–49

Guttmann A, Guttmann RD (1993) Attitudes of healthcare professionals and the public towards the sale of kidneys for transplantation. J Med Ethics 19(3):148–153

Haddow G, Laurie G, Cunningham-Burley S, Hunter KG (2007) Tackling community concerns about commercialisation and genetic research: a modest interdisciplinary proposal. Soc Sci Med 64:272–282

Hann CM (1998) Introduction: the embeddedness of property. In: Hann CM (ed) Property relations: renewing the anthropological tradition. Cambridge University Press, Cambridge, pp 1–47

Hansson SO (2005) Implant ethics. J Med Ethics 31:519–525

Hayden C (2007) Taking as giving: bioscience, exchange, and the politics of benefit-sharing. Soc Stud Sci 37(5):729–758

Hedgecoe AM (2004) Critical bioethics: beyond the social science critique of applied ethics. Bioethics 18(2):120–143

Hetherington K (2004) Secondhandedness: consumption, disposal, and absent presence. Environ Plann D Soc Spaces 22:157–173

Hoeyer K (2002) Conflicting notions of personhood in genetic research. Anthropol Today 18(5):9–13

Hoeyer K (2004) The emergence of an entitlement framework for stored tissue—elements and implications of an escalating conflict in Sweden. Sci Stud 17(2):63–82

Hoeyer K (2006) "Ethics wars": reflections on the antagonism between bioethicists and social science observers of biomedicine. Hum Stud J Philos Soc Sci 29:203–227

Hoeyer K (2007) Person, patent and property: a critique of the commodification hypothesis. BioSocieties 2(3):327–348

Hoeyer K (2010) An anthropological analysis of European Union (EU) health governance as biopolitics: the case of EU tissues and cells directive. Soc Sci Med 70:1867–1873

Hoeyer K, Olofsson B-O, Mörndal T, Lynöe N (2004) Informed consent and biobanks: a population-based study of attitudes towards tissue donation for genetic research. Scand J Public Health 32:224–229

Hoeyer K, Olofsson B-O, Mörndal T, Lynöe N (2005) The ethics of research using biobanks: reason to question the importance attributed informed consent. Arch Intern Med 165:97–100

Holtzclaw D, Toscano N, Eisenlohr L, Callan D (2008) The safety of bone allografts used in dentistry: a review. J Am Dent Assoc 139:1192–1199

Indenrigs- og Sundhedsministeriet (2006) DUT-Høring vedr. forslag til lov om krav til kvalitet og sikkerhed ved håndtering af human væv og celler (vævsloven)

Justiz in Bayern (2008) Entnahme von Zahngold aus der Asche Verstorbener ist strafbar. Press announcement, 29 Jan 2008 (online)

Johnson-Hanks J (2005) When the Future Decides. Curr Anthropol 46(3):363–377

Joyce MJ (2005) Safety and FDA regulations for musculoskeletal allografts: perspective of an orthopaedic surgeon. Clin Orthop Relat Res 435:22–30

Kanal 8 (2009) Wegnahme von Toten-Zahngold ist kein Diebstahl, 17 Feb 2009 (online)

Kellerher S, Christensen K (2005) Baby born after doctor took eggs without consent. The Orange County Register, 19 May 2005 (online)

Kelly S (2003) Public bioethics and publics: consensus, boundaries, and participation in biomedical science policy. Sci Technol Hum Values 28(3):339–364

Knoppers BM, Joly Y, Simard J, Durocher F (2006) The emergence of an ethical duty to disclose genetic research results: international perspectives. Eur J Hum Genet 14:1170–1178

Koch L, Svendsen MN (2005) Providing solutions—defining problems: the imperative of disease prevention in genetic counselling. Soc Sci Med 60(4):823–832

Kranenburg L, Schram A, Zuidema W, Weimar W, Hilhorst M, Hessing E, Passchier J, Busschbach J (2008) Public survey of financial incentives for kidney donation. Nephrol Dial Transplant 23(3):1039–1042

Kungliga Medicinalstyrelsen (1956) Kungliga Medicinalstyrelsens Cirkulär med Råd och Anvisningar Beträffande Organisationen av Blodgivarcentraler. Medicinalstyrelsen, Stockholm.

Last M (1981) The importance of knowing about not knowing. Soc Sci Med 15B:387–392

Lindblom CE (1959) The science of "muddling through". Public Adm Rev 19:79–88

Ling A (2010) UCLA Willed Body Program comes under scrutiny as companies sued for the purchase of body parts. J Law Med Ethics 32:532–534

López J (2004) How sociology can save bioethics … maybe. Soc Health Illn 26(7):875–896

Lundin S (2002) The body is worth investing in. In: Lundin S, Åkesson L (eds) Gene technology and economy. Nordic University Press, Lund, pp 104–116

Macewen W (1881) Observations concerning transplantation of bone. Illustrated by a case of inter-human osseous transplantation, whereby over two-thirds of the shaft of a humurus was restored. Proc R Soc Lond 32:232–247

Mani M (1992) The argument against the unrelated live donor. In: Kjellstrand KM, Dossetor JB (eds) Ethical problems in dialysis and kidney transplantation. Kluwer, Dordrecht, pp 163–181

March JG, Olsen JP (1976) Organizational choice under ambiguity. In: March JG, Olsen JP (eds) Ambiguity and choice in organizations. Universitetsforlaget, Oslo, pp 10–23

Marcus GE (1995) Ethnography in / of the world system: the emergence of multisited ethnography. Annu Rev Anthropol 24:95–117

McNamara IR (2010) Impaction bone grafting in revision hip surgery: past present and future. Cell Tissue Bank 11(1):57–73

Merx H, Dreinhöfer K, Schräder P, Stürmer T, Puhl W, Günther K-P, Brenner H (2003) International variation in hip replacement rates. Ann Rheum Dis 62:222–226

Merz JF, McGee GE, Sankar P (2004) "Iceland Inc."?: on the ethics of commercial population genomics. Soc Sci Med 58(6):1201–1209

Metzler I (2010) Über "Moralapostel" und "smooth operators": Die Praxis der Bioethik im Feld eines österreichischen Biobankenprojekts. In: Griessler E, Rohracher H (eds) Genomforschung—Politik—Gesellschaft. VS Verlag, Wiesbaden

Meyer JW, Rowan B (1991) Institutionalized organizations: formal structure as myth and ceremony. In: Powell WW, DiMaggio Paul J (eds) The new institutionalism in organizational analysis. The University of Chicago Press, Chicago, pp 41–62

Mitchell R, Waldby C (2010) National biobanks: clinical labor, risk production, and the creation of biovalue. Sci Technol Hum Values 35(3):330–355

Moldestad O (2010) Decode Genetics gjenoppstår. Bioteknologinemnda, pp 20–21

Moore W (2005) The knife man. Bantam Press, London

Morgan LM (2009) Icons of life: a cultural history of human embryos. University of California Press, Berkeley
Nilsson A, Rose J (1999) Sweden takes steps to protect tissue banks. Science 286:894
Nordberg K (1998) Folkhemmets Röst: Radion som Folkbildare 1925–1950. Brutus Östlings Bokförlag Sympposium AB, Stockholm
Nuffield Council on Bioethics (2010) Give and take? Human bodies in medicine and research. Nuffield Council on Bioethics, London
Oakley A (2007) Fracture: adventures of a broken body. Policy Press, Bristol
Organization TG (2005) National survey of organ and tissue donation attitudes and behaviors. Division of Transplantation, Health Resources and Services Administration, Rockville
Palmer R (2010) UK may allow payments for gemete donors. Bionews, p 573
Pálsson G (2002a) Medical databases: the Icelandic case. In: Lundin S, Åkesson L (eds) Gene technology and economy. Nordic Academic Press, Lund, pp 22–41
Pálsson G (2002b) The life of family trees and the "book of Icelanders". Med Anthropol 21(3):337–367
Pálsson G, Harðardóttir K (2002) For whom the cell tolls. Curr Anthropol 43(2):271–301
Pálsson G, Rabinow P (2005) The Iceland controversy: reflections on the trans-national market of civic virtue. In: Ong A, Collier S (eds) Global assemblages: technology, politics, and ethics as anthropological problems. Blackwell, Oxford, pp 91–103
Parry B (2004a) Bodily transactions: regulating a new space of flows in "bio-information". In: Verdery K, Humphrey C (eds) Property in question: value transformation in the global economy. Berg, Oxford, pp 29–68
Parry B (2004b) Trading the genome: investigating the commodification of bio-information. Columbia University Press, New York/Chichester/West Sussex
Pottage A (2004) Introduction: the fabrication of persons and things. In: Pottage A, Mundy M (eds) Law, anthropology, and the constitution of the social: making persons and things. Cambridge University Press, Cambridge, pp 1–39
Potts J (2002) At least give the natives glass beads: an examination of the bargain made between Iceland and deCODE genetics with implications for global bioprospecting. V J Law Technol 7(8):1–40
Qvarsell R (1986) Indledning I Framtidens Tjänst. Ur Folkemmets Idéhistoria. Gidlunds, Malmö, pp 9–19
Rabinow P (1999) French DNA: trouble in purgatory. University of Chicago Press, Chicago
Radcliffe-Richards J, Daar AS, Guttmann RD, Hoffenberg R, Kennedy I, Lock M, Seils RA, Tilney N (1998) The case for allowing kidney sales. Lancet 351(9120):1950
Radio Charivari (2007) Verteidiger im Nürnberger Zahngold-Prozess legen Revision ein, 3 Mar 2007 (online)
Rose H (2001) Gendered genetics in Iceland. New Genet Soc 20(2):119–138
Rose H (2003) An ethical dilemma: the rise and fall of UmanGenomics—the model biotech company? Nature 425:123–124
Satel S (2008) Appendix C: public attitudes. In: Satel S (ed) When altruism isn't enough: the case for compensating kidney donors. The AEI Press, Washington, DC, pp 154–157
Scheper-Hughes N (2000) The global traffic in human organs. Curr Anthropol 41(2):191–224
Scheper-Hughes N (2001a) Bodies for sale—whole or in parts. Body Soc 7(2–3):1–8
Scheper-Hughes N (2001b) Commodity fetishism in organ trafficking. Body Soc 7(2–3):31–62
Scheper-Hughes N (2005) The last commodity: post-human ethics and the global traffic in "fresh" organs. In: Ong A, Collier SJ (eds) Global assemblages: technology, politics, and ethics as anthropological problems. Blackwell Publishing, Oxford, pp 145–168
Schicktanz S, Schweda M (2009) "One man's trash is another man's treasure": exploring economic and moral subtexts of the "organ shortage" problem in public views on organ donation. J Med Ethics 35:473–476
Scocozza L (1994) Forskning for Livet. De Medicinske Forskningsetiks Forudsætninger og Praktikker—En Sociologisk Analyse [Research for life. The preconditions and practices of medical research ethics—a sociological analysis]. Akademick Forlag, Copenhagen

Socialdepartementet (1995) Lag om Transplantation m.m. [Law on tranplantations etc.]. Socialdepartementet, Stockholm

Socialdepartementet (2002) Lag om Biobanker i Hälso- och Sjukvården m.m. [Law on biobanks in public healthcare etc.]. Socialdepartementet, Stockholm

Spector M, Kitsuse J (2001) Constructing social problems. Transaction Publishers, New Brunswick/ London

Steinbrook R (2005) Public solicitation of organ donors. N Engl J Med 353(5):441–444

Strathern M (1999) Property, substance and effect: anthropological essays on persons and things. The Athlone Press, London

Taylor JS (2005) Stakes and kidneys: why markets in human body parts are morally imperative. Ashgate, Hampshire

Taylor JS (2008) Donor compensation without exploitation. In: Satel S (ed) When altruism isn't enough: the case for compensating kidney donors. The AEI Press, Washington, DC, pp 50–62

Thacker E (2005) The global genome: biotechnology, politics, and culture. The MIT Press, Cambridge, MA

Thompson M (1979) Rubbish theory: the creation and destruction of value. Oxford University Press, Oxford

Tomford WW (2007) Bone allografts: past, present and future. Cell Tissue Bank 1:105–109

Tutton R (2004) Person, property and gift: exploring languages of tissue donation to biomedical research. In: Tutton R, Corrigan O (eds) Genetic databases: socio-ethical issues in the collection and use of DNA. Routledge, London

Vangsness CT, Wagner PP, Moore TM, Roberts MR (2006) Overview of safety issues concerning the preparation and processing of soft-tissue allografts. Arthroscopy 22(12):1351–1358

Varul MZ (2008) Consuming the campesino: fair trade marketing between recognition and romantic commodification. Cult Stud 22(5):654–679

Whitt LA (1999) Value-bifurcation in bioscience: the rhetoric of research justification. Perspect Sci 7(4):413–446

Wilkinson S (2003) Bodies for sale: ethics and exploitation in the human body trade. Routledge, London

Wilson PD (1947) Experiences with a Bone Bank. Ann Surg 126(6):932–946

Zussman R (2000) The contribution of sociology to medical ethics. Hastings Cent Rep 30(1):7–11

Chapter 5
What Makes "Markets in Body Parts" So Controversial?

If one were a newcomer to the field, I guess it would seem legitimate to ask why some ubjects and some types of exchange cause few public controversies while others instigate hype, hope, and fear as well as legislative reactions. Nevertheless, few scholars explore the differences in the attention given to exchange systems handling, for example, hair, spittle, breast milk, skin, cornea, bone marrow, bone, tendons, blood, plasma, cord stem cell blood, embryos, gametes, organs, heart valves, arteries, muscle tissue, tumors, brain tissue, or dura mater, as they unfold in different arenas regionally, institutionally, and historically.[1] It is obvious from even the most casual observation that some types of ubject exchange, blood samples used in research for example, provoke controversy and stimulate "ethics debates" in some contexts and periods (e.g., genetic research in the 1990s), while exchange of the "same" ubject in other contexts provokes no noticeable reactions (e.g., nutritional research or diagnostic biobanks in the 1980s and 1990s). Also, there are clear differences in the public attention given to, for example, heart transplantation and hair extension, as well as to transplantation of whole hearts versus arteries and heart valves. For me, at least, such differences provoke a basic curiosity, and I believe that such puzzles are important for understanding all the fuss about "markets in human body parts." Since it is clearly not every "body part" generating controversy, we cannot assume that "body parts" hold some inherent, universal, and undeniable moral quality. In fact, as argued in Chaps. 3 and 4, it is never clear what is "part of" a body, and even metal and heat are good candidates. By choosing "ubject" as our analytical term, rather than "body part," we can begin to investigate patterns in the work involved in *making* an ubject into a body part so that it becomes related to a subject (or, alternatively, make it into a plain material resource).

Even if the puzzle about differences in reactions to various ubjects might appear a legitimate scholarly concern, there are good reasons for any scholar wanting to be taken seriously to avoid it. The ethnographic study of various ubject types has distanced itself from its own anthropological past with its grand theoretical projects based on sweeping syntheses across diverse contexts. For reasons I need not rehearse again here, that sort of explicit universal pretension is not appealing anymore.[2] The ethnographic study has become focused on detail, context, and specificity instead. In

K. Hoeyer, *Exchanging Human Bodily Material: Rethinking Bodies and Markets*, DOI 10.1007/978-94-007-5264-1_5, © Springer Science+Business Media Dordrecht 2013

many ways, however, this focus often places ethnographic insights at the margins of other knowledge projects, almost as a counter-thesis on negative terms; we are letting others make the universal claims and limit ourselves to showing how they are *not* valid in our specific location or in relation to our specific topic.[3] It is ironic for a tradition so certain that no universal truths exist, and that all attempts at making them will fail, to be so afraid of formulating ideas on positive terms. With this chapter, I try out some broad comparisons and generalizations in an attempt to understand better when and why public controversies occur in relation to certain ubjects in certain situations. This is a step toward understanding the type of social change stimulated by ubject exchange, and I wish to suggest that it is also a step which deepens our understanding of the stakes in the debate tagged as "markets in human body parts." It is also a dangerous step—not only career-wise (people might think I am trying to propagate a grand theory)—because it implies moving between two traditions (an old comparative one and a contemporary ethnographic one) as well as between two dimensions (a material one and a semantic one) with different parameters of comparison (ubject or context).

As I try to reinvigorate comparisons, I must also make it clear that I do so from a very different ontological and epistemological set of premises than the classical theories. My theorizing is case specific, as I made clear already in Chap. 1. While I do wish to reach positive conclusions, I have no universal pretentions. Still, many ethnographers will find that what they usually look for as the treasured qualities in a text do not fit the comparative mode of inquiry; there is no "thickness" to the description, no deep sense of context, no detailed ethnography. I am searching for that which is not revealed through attention to detail, that which we tend to lose sight of when seeking to explain phenomena through meticulous analysis of specific historic settings. I view the comparative exercise as complementary to detailed ethnography, and I think that we understand better the importance of context and detail if occasionally we make broad and sweeping comparisons too. Some "local ethnographies" have drawn far too sweeping conclusions about the "nature of the body" or the workings of a "market," simply for lack of awareness of their own specificity.

I have already drawn on Marilyn Strathern's notion of partial connections to describe my own theorizing as a shifting set of choices about what to foreground and compare (Chap. 1), as well as the type of theorizing people undertake when making sense of the ubjects they face (Chap. 3). The notion of partial connections also helps explain what this chapter seeks to do. Either we look at the same ubject in different contexts or different ubjects in similar contexts. "Context" can be defined, for example, spatially, nationally, culturally, or institutionally, and differences in context can reflect changes over time or between places. When I suggest exploring when and why certain ubjects cause controversy in certain contexts, I move back and forth between these modes of inquiry, which foregrounds different aspects of ubject exchange and in effect creates different objects of knowledge by way of making shifting partial connections. My balancing of emphasis on ubject and context also relates to another recurrent theme of this book, which has to do with the balancing of material and semantic dimensions of exchange. When foregrounding the ubject at the expense of context, I will have a greater chance of exploring the material agency of the ubject, while foregrounding of context will tend to emphasize the attribution of meaning. As I try to move between these poles, I traverse a wetland in which any claim can be erased through the making of another set of connections. It should be obvious, therefore, that I am not making a set of definitive claims about the (everlasting) nature of ubjects; rather, I play with comparisons to move beyond the naturalized claims already inherent in the typical positions taken by market proponents and opponents.

I begin with a presentation of a list of ubjects, mainly to expand the examples that we tend to include in our contemplation of ubject exchange but also to return to the basic issue of what makes an ubject into an ubject in my sense of the word. I then propose a personal interpretation of why certain ubjects cause controversy in certain situations. My claim is that controversies reflect attempts of demarcating the worthy from the unworthy and that this exercise is particularly important when people are dependent on the care of others. When making such distinctions, we draw on purified representations of the world. However, we cannot escape hybridity and we should embrace also the ambiguity embedded in our phenomenological experience to refine our understanding of ubject exchanges. I therefore in the subsequent section offer some reflections on the study of ubjects as a very intimate research practice before summarizing what I think we might learn from this type of comparison.

Typology as Entry Point

If we accept viewing bodies as material flows rather than entities, as I suggested in Chap. 3, it logically follows that no flow consists of predefined and well-delineated entities. Many flows have been hypostatized and given a name—teeth, hair, nails, vomit, spittle, etc.—but any ubject can enfold another (as teeth enfold blood, which enfolds cells, which enfold DNA). This is why we should think of an ubject not as an entity with a name but more like a point in time between being part and not part of a body. What type of part, what name it will acquire, depends on the mode of inquiry and the interests in the ubject.[4] This understanding stands in sharp contrast to a positivist science pursuing classification of ostensibly preexisting material entities in typologies such as those developed by Carl Linnaeus (later ennobled as Carl von Linné). Typologies—these favorite tools for positivist science—are interesting because as works of purification, they bring attention to hybrids. And they do indeed facilitate contemplation of differences and similarities across various ubject types— if only you do not reify the "types." In Table 5.1, I have therefore played around with hypostatized and named ubjects.

The table illustrates some of the breadth of what I think the term ubject allows us to capture, but remember that it is not intended to be an exhaustive list; indeed, there is no such thing as an exhaustive list.

To organize the list, I have had to draw on various distinctions. I now briefly discuss these distinctions and the delineations of the meanings of the term ubject that the list implicitly conveys. I do not subscribe to these distinctions at an ontological level. On the contrary, by fitting onto the same list, for example, both "biological" and "nonbiological" ubjects, I wish to allude to the way all flows are precariously related to the body irrespective of such distinctions having a particular currency today. The idea of the biological has a particular historical genealogy reflecting changing ideas about the nature of "life," and it cannot be settled from first principles which parts of the material world are biological or nonbiological.[5] It depends on what is seen as part of what, and again this exercise of definition is what we engage when dealing with ubjects. Individuals might use notions of biology to determine some ubjects as alien—for example, bullets, shrapnel, or other "leftovers" from war injuries—but many small objects enter bodies and become encapsulated without being viewed as "alien" or particularly "nonbiological." Nevertheless, by employing common distinctions as those in the table, we acquire enhanced awareness of the shortcomings of the categories otherwise employed matter-of-factly to set aside, for example, tissue from devices, medical from nonmedical, research frontier practices from mundane everyday flows. Classical anthropology typically studied the ubjects emerging through daily self-practices, while contemporary studies tend to focus on ubjects generated through more complex medical and nonmedical interventions. The latter always involves other users than the "donor" and a set of politically mediated institutions. This institutional change is probably important if we wish to understand why social science inquiry is drawn to them.[6] Still, all the various types of ubjects in the typology share more features than usually assumed, I believe, because they all interact with the fruitful darkness of in-betweenness as well as with the work we undertake to delineate the self (cf. Chap. 3).

Why (Not) Include…

A list of this type can also be useful in elucidating some of the underlying assumptions of the typologist.[7] Why, for example, have I not included drugs on it? In Andrew Lakoff's work on perceptions of psychiatric research in Argentina, there is an interesting discussion of psychiatric drugs as being part of, or external to, the person:[8] on the one hand, the biomedical thinking on which medication is based is a materialist perception of the person as biochemistry, while on the other hand, medication presumes an external person wanting to be biochemically different. Drugs are therefore contested as part of/not part of the person, and as such they could be seen as ubjects. What legitimizes medication is an autonomous decision maker (not unlike a consumer) construed as residing outside that very body, which is simultaneously seen as pure materiality. This resembles the paradoxical way transplant tissues are construed as "just material" elements of a soulless, secular body that can be accessed legitimately only through informed consent granted by an autonomous agent somehow external to the body. Since drugs rarely feature in the debate I address in this book, however, I use this example merely to specify the choices I—as a typologist—have made and to indicate how the analysis could have taken other routes if addressing other cases.

Table 5.1 Ubject types leaving the body

Mode of production	Biological or nonbiological	Distinguishing features	Examples (not exhaustive)
Bodily self-practices	Biological	Mundane bodily emissions	Urine Feces Sweat Saliva Menstrual blood Sperm Breast milk Hair Nails Teeth Exfoliation Dust (DNA traces)
		Secretions associated with malign circumstances	Necrosis Inflammatory secretions Vomit Cadaver
	Nonbiological	Cosmetics	Piercing Artificial nails Artificial hair
		Injuries	Bullets Shrapnel
Medical interventions	Biological	Treatment (mostly related to transplantation medicine)	*Low/medium levels of processing* Breast milk Bone Tendons Cartilage Teeth Skin Dura mater (meninges of the brain) Cornea Blood and plasma Urine (for hormone extraction) Bone marrow Organs Heart valves Arteries Gametes Fat
			High levels of processing Tissue engineering Advanced blood products
			Promissory technologies Various yet-to-be stem cell technologies using embryos, umbilical cord blood, adult stem cells, etc. Gene therapy

(continued)

Table 5.1   (continued)

Mode of production	Biological or nonbiological	Distinguishing features	Examples (not exhaustive)
		Diagnostics	Tumors Blood Mouth swabs
		Research	*Biobanks (cohorts or disease specific)* Blood, mouth swabs, urine, tumors, teeth, muscle, bone, etc.
			Cell lines Cell lines as research tools Cell lines as tools of production Feeder material for cell lines
		Education	Specimen collections Corpses
	Nonbiological	Active devices	Pacemakers Implantable cardioverter-defibrillator (ICD)
		Passive devices	Prostheses Screws, etc. Tooth fillings (gold, mercury, etc.)
Nonmedical interventions	Biological	Knowledge production (identity and history)	DNA (for ancestry research or ancestry testing) Bone (archaeological/physical anthropological research)
		Forensics	DNA Pathological evidence
		Entertainment/ education	Plastinated bodies (Body Worlds)
		Aesthetic uses	DNA sequencing on textiles (GATGee) Art work Implants
		Memorial uses	LifeGem jewelry Hair jewelry
	Nonbiological	Surveillance/identification purposes	GPS implants Identity chips for soldiers

Probably some readers will also have found a different set of phenomena missing on the list, namely, the various material representations arising from interaction with the body. I am thinking of photos, sculptures, death masks, as well as all the many technologically mediated representations that suffuse contemporary medicine in the form of visual representations (e.g., scans and x-rays), numerical representations (e.g., blood pressure counts, and sugar or lipid levels), or other types of diagnostic test results (e.g., genetic, bacterial, or viral tests).[9] Since they are ambiguously related to the body, they could have been discussed as ubjects, but while they do indeed interact with how we come to understand ourselves, their mode of exchange is rarely discussed along the lines of "a market in human body parts" and therefore fall out of my field of interest.

The typology facilitates not only contemplation of distinctions and ubject definitions. By comparing ubjects usually seen as unrelated, we can begin an exploration of why some ubjects attract so much popular and scholarly attention while others continue to live relatively private lives. This is my real concern and the topic the rest of this chapter explores. Just a brief glance at the list at least facilitates debunking some simple answers that might otherwise have had appeal. Policy attention and public controversy do not depend, for example, on whether there are few or many of a certain type of ubject in circulation. The bodily self-practices generally generate more ubjects than the mediated practices but are mostly seen as less controversial in contemporary Western societies.[10] Even if limiting ourselves only to the ubjects generated through medical practices, it is obvious that it is not the volume of ubjects that determine the level of public attention: embryonic stem cells are highly contested unlike the much more common bone grafts. Another easy answer we can dispense with is that novel ubjects should be more controversial than old ubject types. Blood, for example, can suddenly become contested in tandem with changes in the exchange form and by way of new uses or user groups, as described in the preceding chapter.

If we wish to understand why some ubjects live such turbulent lives while others, in certain periods at least, appear too mundane to attract any attention, I would suggest considering another set of factors. The point is not to present a definitive set of determinants for controversy. Any ubject embodies, so to speak, a potential for controversy. My point with this exercise and the generation of a list is to contemplate the issues at stake in the controversies currently tagged as "markets in human body parts" and thereby deepen the analysis of donor interests from Chap. 4. I think we should take these issues seriously because by exchanging ubjects we potentially interact with very basic understandings of who and what a person is and how persons should interact with each other.

Characterizing Controversies

The first and most obvious thing we should notice is that the public needs to know about a practice for it to cause controversy. This almost goes without saying, but it needs to be said because some scientists seem to think that what they do in the lab does not provoke people, while in fact many people probably would be provoked if only they knew. Second, and related to knowing too, the relationship between the body in whom the ubject originated and the group of ubject *users* is of key importance. As long as an ubject stays within the boundaries of a defined user group having similar interests in it, it is less likely to cause controversy than when multiple user groups are involved.[11] Think of the difference between LifeGem jewelry, where a company makes an artificial diamond out of the ashes of a deceased relative, on the one hand, and the exhibitions of plastinated bodies of Body Worlds which are exposed to the public, on the other.[12] In some respects, these are similar practices, transforming a corpse into an aesthetic object, but the range of user groups is very different and the level of uncertainty about the origin of the ubject greater with the plastinated bodies. People commenting on LifeGem generally see it as a curiosity, while Body Worlds is widely debated, admired, and contested. Third, and related to user groups, are the material implications of ubject transfers and the multiple meanings they carry. It is generally more controversial when the material implications are seen as related to survival, creation of new biographical lives, or of new life forms than when ubjects are seen as merely restoring lost capabilities (such as bone transplants). Fourth, ubjects that at a phenomenological level cause pain or interact with the sense of control will be more controversial for the involved persons than ubjects without such effects, as argued in Chap. 3. Fifth, institutional and cultural context seems important in the sense that embryos do not connect with the same controversies in, for example, the USA or Denmark (where abortion is uncontroversial).

It is tempting to say *medicine* is an institution causing tensions around ubjects that, for example, aesthetic uses do not. To some extent, medicine really is a particular institutional context in which ubjects take on particular meanings. However, many artists have used ubjects to provoke (and managed to achieve their aim)[13] and many medical practices are seen as mundane (e.g., diagnostic blood tests). Hence, it is not medicine as such that produces so many ubject-related tensions. As with the "medical realm," it is tempting to argue that the "market" makes an ubject controversial by activating the moral paradigm of market thinking whereby the ubject becomes more subject-like. Indeed, the previous chapter showed how controversy can result from introduction of commercial incentives into ubject exchange. However, Chap. 2 made clear that there is no such thing as a "market" which can be clearly identified and ascribed explanatory power. We need to understand better what it is about some types of exchange that activates concern. Beginning with a reinterpretation of what "markets" do and moving onto a reinterpretation of what is at stake in "medicine," I now present my interpretation of what it is that typically makes ubject exchange in the realm of medicine and "market-like" transactions so controversial. This interpretation takes us back to the historical rise of particular institutions, modes of thinking, and exchange practices begun in Chap. 2.

Reinterpreting What the "Market" Does

Typically, associations to "markets" are stimulated by the presence of money, commercial operations, and monetary incentives, and I wish to suggest that any of these potentially activate "the market" as a marker for a boundary that used to demarcate the sacred from the profane. They acquire this function because the "market" indicates an exchange form which is supposed to be ruled by self-interest rather than expressions of care; they mark a separation of a realm of ends (subjects) from a realm of means (objects). The depth of this claim must be understood in light of the rise of a secular society and the societal changes outlined in Chap. 2. In her important work about the founding of anthropology in France in the late nineteenth century, Jennifer Michael Hecht describes how intertwined the founding of the study of *anthropos*—man—was with materialism, scientism, rationalism, and the pursuit of a secular order. The atheist project was related to freethinking and opposition to the conservative role of the church in the second empire.

Hecht points to one of the central problems for the proponents of a secular society: What to do with the soulless body? The freethinkers of the time were deeply engaged in legalizing civil funerals, but the problem went further than simply how to dispose of the body without church liturgy. How should one *make sense* of the body and its relation to the person when it was no longer a mortal frame for an eternal soul? Hecht suggests that the founding of the Society of Mutual Autopsy served this purpose for these forerunners of the secular order. They decided to donate their bodies to each other and through mutual dissection advance the main tenets of what Hecht provocatively calls their *faith* in science and progress. A utilitarian view of the body emerged and formed a belief system in which the dominant distinction became the one between the useful and the useless.[14] For the useless, they could feel nothing but disdain, while the useful (in particular the brain) was to be conserved, exhibited, and used for science. In this way, the distinction between waste and usefulness served to delineate the good from the bad, the worthy from the unworthy elements of their body. Importantly, however, it was not any use that provided meaning. While explicitly materialist and empiricist, these freethinkers were also humanists. They wanted progress for humankind, and putting the cadaver to use in a meaningful manner was about serving a meaningful goal. Their world was not a plain material continuum—the ubjects they donated were to serve a proper purpose: the well-being of human beings. Through association with a proper purpose, the donated brains could remain attached to a "human" world.

The French experience is special in many respects, but the shift toward secular societies in which the body comes to be seen as a *useful* material resource is more general and so is the shift toward thinking that a person should decide over the fate of his or her own body as a material resource.[15] The rise of a secular order interacts with the emergence of a belief system in which rationalism lends meaning to that which is ambiguously related to the human subject. When faced with the task of handling that which is bound to fall "beyond the limit—*cadere*," in Kristeva's sense, people could now revert to a rational utilitarian belief system and ascribe new meaning to their ubjects. Even religious people were now provided with an additional mode of assessing bodies. In his seminal work on religion—which was part of the secularization described by Hecht—Emile Durkheim pointed out that every society needs systems for classification, an order of things:

> All known religious beliefs, whether simple or complex, present one common characteristic: they presuppose a classification of all the things, real and ideal, of which men think, into two classes or opposed groups, generally designated by two distinct terms which are translated well enough by the words *profane* and *sacred*.[16]

The parallel to Latour's notion of purification work is obvious. Durkheim differs from Latour, however, when he elsewhere describes the sacred and the profane as mutually exclusive, as "absolute" separated by an "abyss."[17] Yet he also acknowledged that relations of power and deference spilled over into the handling of the sacred and that the

> sacred world is inclined, as it were, to spread itself into this same profane world which it excludes elsewhere… This is why it is necessary to keep them at a distance from one another and to create a sort of vacuum between them.[18]

As remarked in Chap. 3, Durkheim also noted that bodily substances such as blood seemed to hold a particular potential for making things sacred: pouring blood over an object would set it aside from the rest of the material world. With this in mind, and considering how Hecht insists on seeing this secular approach to the world as a belief system in its own right, think about the way in which people in Chap. 4 praised usefulness saying, for example:

- Really, for me it's just a piece of waste, right, so I really don't care. If it can be put to use, it's fine.
- All these immense amounts of heat and they just throw it out—to no use [indignant]! You might as well use it!
- If it's to be of any use, it shouldn't be too costly.

while simultaneously disliking the thought of giving the bone to the dog or selling the heat commercially. *They employ the notion of usefulness to provide ubjects with meaning; they draw upon a rationalist world order to make sense of ubjects.* As the system of classification delivered by the church has lost authority, new modes of orderings are introduced even if they still serve to *set aside the worthy from the unworthy.* In her analysis of the rise of a life insurance industry in the nineteenth century, Viviana Zelizer also uses the distinction between sacred and profane to explain the initial resistance in several countries to commercial markets based on calculation of life expectancy. To count people was by some regarded as sacrilegious to the extent that they even opposed censuses. Zelizer writes that sacred things "are distinguished by the fact that men will not treat them in a calculating, utilitarian manner."[19] My point, however, is that we are dealing with what is no longer about the sacred but exactly a *secular* heir to this distinction.

Hecht also points out how the rise of secularism stimulated a new set of rituals to feign a boundary between the human and nonhuman world. Following Constantina Nadia Seremetakis, we can talk about this as an ongoing ritualization (rather than a set of fixed rituals).[20] As societies produce new ways of ritualizing the passage from the sacred to the profane, the human to pure materiality, informed consent has now become a central feature of many such ritualizations. Why is it necessary to do all of this? Perhaps it is not, and certainly some claim that all of this is absurd, but very many people seem to have a basic phenomenological experience of bodies as animated, willful persons constituting more than material components of the same worth and standing as any other part of the material world. Bodies are different for them than stones, soil, and water. They need a marker to separate the soulless body from the rest of the material world, as the sacred for millennia has been separated from the profane.

Market thinking is a close relative of the rationalism characterizing secular societies. However, the equalizing effect of the presence of money creates an expectation of total fungibility: any of the elements—money, for-profit organizations, or monetary incentives—activating notions of markets thereby become markers of the mundane, material, dispensable, and commensurate. For people who want to retain a boundary between the bodies of willful beings and the rest of the world, or between the worthy and the unworthy, "market-like" exchange forms of ubjects become abject. The dismal reference to commodification should therefore be seen as indicating a secular belief system in which ambiguous ubjects are ascribed particular types of value. The idea of a "market" can therefore be used by a secular society to distinguish between the worthy and the unworthy, the truly deserving and the plain raw materials used to fulfill their (legitimate) desires.

If we accept this interpretation, we might also better understand what for many Europeans appears as plainly primitive and uncivilized about the USA: the much greater propensity for applying market thinking to almost anything, including ubject exchange. Clearly, a majority of the market proponents discussed in this book are American, and the USA is known as the place where gametes and plasma are treated in ways most closely resembling a commodity form. It is even common for American proponents of organ markets to refer to the sale of plasma and gametes as utterly unproblematic and uncontroversial (apparently not aware of the disgust it can awake in Europe) and to use this analogy in support of kidney sale. One reason for this difference might be that the USA is, by European standards, a quite religious society in which a majority believes in the existence of the soul and a divine order—in contrast to the more secular Nordic countries where the opposition to ubject trade seems to be fiercest.[21] One possible understanding of this difference would be that with religion ever present, there is not the same need in the USA for a successor to the distinction between the sacred and the profane as the one felt in the Nordic countries: respect for the person can be attached to the soul (guarded by God) while the body becomes entangled in mundane exchanges along with the rest of the material world.[22]

For some observers, boundary setting practices appear irrational, whether conducted secularly or religiously. Philosopher Stephen Wilkinson, for example, knows his Kant, and yet when he thinks carefully about the commoditization arguments, they all seem to dissolve and appear unfounded. Dedicated to rationalism, utility, progress, and humanism—the central values of the secular society—he concludes that there are no arguments from first principles against a market in organs.[23] I agree, but, frankly, I find it more surprising that anybody could have assumed that such an argument could have delivered ultimate and definitive boundaries designating what can be exchanged and how. The worldviews we articulate employ distinctions and purified assumptions, but the lifeworlds we actually live in are full of hybrids evading final classification. The question is whether we should revert only to worldviews and deductive thinking when seeking to understand what is at stake in recurring controversies. Clearly, philosophy of the type Wilkinson exhilarates in can help overcome many prejudices, but there is wisdom also in the phenomenological experience of the lifeworld. I am not thinking of "wisdom of repugnance"—a so-called yuck factor—in the sense suggested by Leon Kass.[24] What I have in mind is what we learn from placing ourselves among the people who are engaged in the everyday activities that evade easy characterization and evaluation. By sharing their entanglement with the morally ambiguous, we understand better the pros and cons. The resistance to market-like exchange of ubjects informs us about values attached to the boundary between the world of the worthy and deserving and the commensurate, unworthy, mundane world of materiality. Today, designation of worthiness can find no socially robust grounding in theology and not even in secular humanism (not all humans are deemed worthy ends, while some animals and material objects are); it has no grounding outside the willful beings wanting to protect particular aspects of their world. To achieve that protection, they continuously create and re-create a functional and flexible distinction between the more worthy ends and the lesser worthy resources placed in the service of these ends.[25]

There is no such thing as a "market" out there, and it is not "markets" that produce controversies around ubjects. But in some societies, there are features of exchange that are understood as market-like and which with differing force activate the boundary between the deserving ends and the plain resources meant to serve them. Some people feel very strongly about this boundary. This boundary is a central element of a biopolitical landscape designating some elements of the world as fair game and others as protected and special. Furthermore, the same features of exchange relations indicate a type of exchange in which self-interest and desire fulfillment are legitimate. When exchanging what people do view as commensurate, mundane, and fully fungible, these features are just fine, even comforting. But in situations where they are dependent on the care of others—where they do not want the surgeon to "take any more than supposed to" (Chap. 4)—these features appear less appealing.

Reinterpreting the "Medical Realm"

The same longing for trust is also what we should appreciate when re-approaching the question of the "medical realm" as a place where ubjects are likely to cause controversy and drama.[26] I wish to suggest that it is not medical institutions as such that make ubjects special to many people; rather, it is the situations where people are dependent on the care of others and it is the effect of certain experiments conducted by medical researchers on our public imagination of humanness and selfhood. The first point is obvious, but rarely appreciated: ubjects matter when procured, exchanged, and used at a time when the very opening of the body is associated with personal risk. Patients about to undergo surgery need to trust the person in whose hands they are about to place their lives. As a patient, you do not want the doctor to be distracted by the wrong incentives. But even in the general public and in cases where no patient's life is at risk, the medical realm has become a prime producer of controversial ubjects.

The technologies primarily receiving ethnographic attention in relation to the literature on "tissue economies" illustrate this further. There is a propensity in medicine for developing technologies that change how human beings are done. Of course, it does not make sense to talk about "medicine" in the singular as a defined entity, just as there is no "market" out there. Even controversial subdisciplines such as genetics tend to take on different national forms, as argued by Karen-Sue Taussig, among others.[27] But my point is that medical research has become a prime arena for reconfiguring biological life. As a consequence, we might think of it as a laboratory not only for developing medical techniques but also for rethinking limits to humanness. As elements in such reconfigurations, ubjects become material artifacts to which a range of associations can be affixed. Many of the ubjects exchanged in the medical realm(s) are used in research that proclaims to fulfill the vision of progress that was propagated by the secular freethinkers of late-nineteenth-century France but which continues to scare many people more comfortable with life as they know it. Probably, any innovation interacting with how human bodies are done will provoke both hope and fear (to use the terms Michael Mulkay employed to describe the embryo debate in Great Britain),[28] and the ubjects used in furthering these agendas are part of promoting a sense of fruitful darkness from which anything can arise. What we see is medicine providing a set of arenas in which people might contemplate their boundaries and material constitution. Previously and elsewhere, religion has raised such questions in relation to ubjects now generally seen as utterly mundane and uninteresting, such as hair and nails, as pointed out by Bynum.

Summing Up the Argument

I have argued that controversies relate to attempts of demarcating the worthy from the unworthy and that this exercise is particularly pertinent for people when expressions of respects for their worth are dependent on the care of others, of strangers, of people or systems over whom we have no authority. In the previous chapter, I suggested that from the perspective of potential donors, the appeal of commercial ubject exchange will reflect available alternatives and the relative sense of control. By turning my attention to public controversies, I have pointed to societal interests beyond the concerns of the individual donor. Controversies about "markets in human body parts" have become instances where society establishes lines of demarcation of a more general nature. The stakes are high when we decide who and what comes to be placed on the worthy or the unworthy side of the equation, and my point is that we should appreciate these stakes rather than either applying market thinking to ubjects in a manner that designates them as plain material resources or applying universal references to human dignity and assume that no material passing through the bodily space can ever again be exchanged in a manner that acknowledges its monetary worth.

Stated briefly, we can conclude that the reasons for some ubjects being more likely to cause controversy than others include interference with a need for setting a boundary between worthy ends and plain material resources, demarcation of realms in which the weak can rely on other's care, the user's relationship to the person in whom the ubject originates, phenomenological experiences of pain and the sense of control, cultural context and the partial connections that can be made, material implications of ubject transfers, and public communication and awareness of ubject exchange. In addition to these points, I have pointed to waste as a crosscutting category helping to distinguish the useful from the useless and infusing bodies with meaning in secular, "rational" societies and thereby softening controversies. Waste is also a marker of harmlessness in a given intervention (as when the surgeon explains "You can't use this. But we can. Is it alright if we keep it?") and thereby interacts with the sense of control.[29]

Ethnographers are in a privileged position of moving beyond the purified representations of the world used in the power games of demarcation. I now turn to some reflections on the study of ubjects as a very intimate research practice which are meant to stimulate the ethnographic entanglement with the inherently ambiguous.

Reflections on the Study of Ubjects as an Intimate Research Practice

Studying ubjects is in many ways a macabre occupation. When hearing about my research interests, friends, family, and colleagues often volunteer stories of their own or forward links and clippings about various forms of ubject exchange. Now, I wish to discuss a certain ambivalence which surrounds my at times almost morbid enterprise. My interest lies in the ways a study of ubjects arouses fervent emotions and yet remains strangely cynical when turned into a calm, everyday practice. I believe that by staying alert to this type of ambivalence, we also learn something important about the ambiguity characterizing ubjects. Our basic phenomenological experience as researchers as well as persons (and ubject producers) can help us navigate the wetlands of ubjects.

Most people working with or writing about ubjects move back and forth between different emotional states. Sometimes ubjects appear as mundane material objects; sometimes they seem very human and special. I first really got aware of this when I was studying the Swedish biobank described in the previous chapter. One morning, I was shown around the storage facilities at the Medical Biobank. We had been walking down the rows of freezers, and I was told about the storage principles and the security arrangements surrounding the tissue collection. My mood was as calm as the room: I took notes, looked down at individual blood samples, and noted that "each line in a rack comes from one person." As my guide and I went out the door and he turned off the light, he mentioned, "So this is where we keep the blood of 85,000 people." His words suddenly conjured up an altogether new image in my mind, and I felt that I had visited a kind of graveyard or a sanctuary in which 85,000 people were kept. It was an important moment because I suddenly realized the extent to which the tissue, the milliliters of blood, stored in the Biobank is open for many interpretations, and each produces a range of emotional effects, even for the same person. In this light, the paradoxical behavior of people who, on the one hand, felt strongly about what their blood was used for but, on the other hand, did not show any interest in informed consent sheets made more sense. Chapter 2 presented several examples of health professionals moving back and forth between different conceptualizations when working with ubjects, and Adam Benchard, Kathryn Ehrich, Cecily Palmer, and Bronwyn Parry have all described how the presence of (what I call) ubjects performs this ambiguous work on the people handling them.[30] In line with Durkheim, we might say that it takes a strong belief system to stabilize an official reading of the ubject—an ordering of the world—to convince people that an ubject *really* is just an object (or an enduring part of a subject for that matter). The debate about "markets in human body parts" exemplifies this kind of purification work, but the everyday phenomenological experience of handling ubjects continuously effaces the temporary certainty produced.

I think we should use this recurrent feeling of uncertainty to understand better the stakes in ubject exchange. One of the most reflective scholars trying to move beyond the gift/commodity dichotomy is geographer Bronwyn Parry. In a particularly interesting piece, she reflects on her own emotional responses to stored brains during fieldwork in a brain bank.[31] She had made jokes about brains, even seen them as hilariously funny objects, and moved around in the bank without sensing any particular emotional reactions, when suddenly confronted with the brain of her favorite author, Iris Murdoch. Just then, right there before her, the ubject changed its very being. She felt crushingly intimate, far too intimate in fact, and she wanted to protect the brain from harm and from the public gaze. She even felt pity for Iris there in the bucket. Parry's experience holds a lot of similarities to what people express in relation to known rituals in the anthropological literature of second burial.[32] Reading the article, I, too, intuitively knew what she meant. I had been there, just as I guess most readers have had experiences of this kind of unexpected intimacy when confronted with the material presence of something ambiguously related to the people we associate with them, for example, a sample of hair from a missed one or an intimate scent in an unforeseen spot. In some encounters, what used to be an object turns into an evocative ubject demanding our protection and care. It is our phenomenological presence and our sense of will (which are more than intellectual schema) that overrule the "classification of all the things, real and ideal, of which men think" (Durkheim's phrase). We should use these experiences to acknowledge that it is not clear what ubjects are; they embody multiple ontologies, and they work on us and our will as much as we work on and with them to fulfill our desires.

Parry describes her experience as "uncanny." What is it that makes something uncanny and what does it imply? In Chap. 1, I mentioned Freud's classical essay on the uncanny, and when I now return to this, it is because I think it might help us to reflect on how to draw on our own emotional responses as a reservoir of knowledge about ubjects. Freud, unsurprisingly, suggests that events are uncanny if they reintroduce something previously repressed. Frankly, I am not interested in the -cognitive or psychoanalytical dimension of uncanniness. What interests me is when he, with reference to Schelling, suggests thinking of "the uncanny as something which ought to have remained hidden but has come to light."[33] Freud argues that "intellectual uncertainty has nothing to do with the effect." The uncanny is not limited to intellectuality, but uncertainty is an important part of creating the feeling. In the uncanny moment, we are in a state of not knowing what to make of the feelings produced by the ubject. Our categories do not apply. I believe we can use these moments to get closer to that which "has come to light." Usually, our texts, simply by virtue of being *texts*, express our worldviews, that is, our spoken categories, perceptions, and explicit value commitments. The moment of the uncanny, in contrast, is a moment of phenomenological presence. It is a moment of lifeworld immersion. In this lifeworld, ubjects are fundamentally unstable: they are *not* simply "worthy" repositories of dignity or mundane, plain materiality. They are betwixt and between.

Laughter

When we take our lifeworlds seriously, we engage the continual flow of hybrids. Often during my talks, people have giggled and laughed, just like my informants sometimes laughed when reflecting on the recirculation of heat.[34] Simultaneously, however, the same topic can also arouse an almost sinister sincerity, as when they come to agree with the ethics council that restricted recycling of heat is a preferable option. We seem to laugh in reaction to absurdity, but laughter also provides a para-doxical opportunity to both connect with and distance ourselves from that which does not fit the categories of our worldview. The moment of laughter is one in which we embrace the ambiguity of that which simply does not fit. To some extent, we laugh of the purification efforts of others; and yet, when I invite people to express their views and make their own judgments, they do so with sincerity. They seek to purify because it really *does* matter to them what is and is not part of a person. The purification work concerns more than poorly matching *words*; the categories embody a phenomenological dimension relating to categories of experience—the material experience of objects and subjects and of our attempts of delineating the two. Furthermore, I think of the recurrence of laughter as connecting us not only to the topic as such but also to one another. It connects us in a common pursuit of delineat-ing that which does not fit, but I am also thinking of the *type* of laughter as related to a particular form of absurdity which unites people in recognition of their shared corporality.[35] When we laugh of the image of a dog lying on a carpet with its owner's hip in the mouth or of the predicament faced by crematoria trying to main-tain difference between various forms of heat, it reflects a confrontation with the grotesque. I am inspired here by literary scholar Mikhail Bakhtin's reading of Rabelais' work.[36] Bakhtin points out how the renaissance author Rabelais provoked laughter through grotesque scenes that explicitly exposed transgressions of body boundaries: urine, feces, and enormous mouths. Bakhtin suggests that laughter arose because the audience was reminded of humanity's shared materiality—our own banal corporality as it were or you might say the ridiculous aspects of trying to lift humanity out of this shared materiality. It was a laughter uniting people around something common to everyone. Bakhtin suggests that Rabelais lost his audience when the bodily banal came to be seen as plain vulgarity. In our dealings with ubjects, however, it seems we still face the grotesqueness of shared corporality in ways capable of provoking a good—shared—laughter.

This book is of course affiliated with the empiricist, secular, rational, materialist tradition described by Hecht. Disciplinary affiliation always provides a mindset with which to approach one's object of study, and every study contains its own categorization of the real and unreal. I believe, however, that the somewhat cynical, calm, and collected attitude that facilitates the creation of a table of ubjects such as the one in the beginning of this chapter, as well as the contemplation of the "meaning" of ubject narratives presented in the previous chapter, must be countered by not only playfulness but also a bit of the emotional density of the lifeworld to arrive at anything more than a recirculation of known elements of cultural narratives. By engaging our own responses to the "uncanny"—as entry points to what I call the wetlands of ambiguity—I think we appreciate better what it is that people are trying to settle in ubject debates.

"Ubjectology" as a Vocation

Very basic debates about what is part of whom, for how long, and with which implications are continuously reopened in different registers over time. At some points, they appear as religious debates, then as secular debates, and recently within a particular secular frame as economic debates. Instead of trying to close the case with answers to what is (part of) a human being, I think we should appreciate the fact that we do not seem to be able to reach any permanent agreement. This would allow us to devote more attention to who (and what) benefits and who (and what) loses with one or the other classificatory approach. We might also accept that there is a lot of similarity between the secular attempts and the religious attempts of delineating a person materially. The undefined keeps producing new tentative answers, none of which will satisfy the subsequent generation.

As already suggested in Chap. 2, I see in the intense debates about what can be exchanged, and how, a society delineating itself—much like Laporte suggested a person delineates himself or herself through continuous acts of purification. The many attempts at controlling the conditions of ubject exchange indicate a society which, to borrow from Durkheim, "confidently affirms itself and ardently presses on toward the realization of the ends which it pursues."[37] This chapter set out to explore why some ubjects leave the body quietly and without much notice while others become centerpieces of controversy. Controversies evolve in relation to the making of significant boundaries. I have played around with typologies to stimulate reflections on what counts as an ubject as well as what characterizes various ubject controversies. This return to one of the favorite preoccupations of the empiricist tradition from which the social sciences developed as a secular vocation also served as a (slightly ironic) stepping stone toward an argument about what seems to be at stake in ubject controversies. But I have also suggested using the phenomenological experience of the lifeworld as a resource for appreciating a very basic ambiguity surrounding ubjects.[38] Engagement with the *presence* of ubjects implies confrontation with both material and semantic aspects of the ubject. Different ubjects will raise different associations depending on cultural context and the semantic landscape facilitating particular partial connections. When we consider local reactions and controversies surrounding ubjects, the phenomenological experience of pain as well as the personal feeling of control are of central importance.

The point of this whole exercise is that *if* we see ubject controversies as indicative of values at stake for the involved actors, and *if* we wish to respect these values, *then* we need to pay particular attention—not so much to the supposed ontological status of the ubject, as often seen in the debate, but—to the instances when ubject exchange interacts with people's sense of control, to the hopes they have of being met with care, and to the way in which the categorization of the worthy and deserving always implies designating something as plain and legitimate resources, as fair game.

Something very basic and important is at stake when we are exchanging ubjects. Bynum's work on the rise and spread of the doctrine of material continuity as a dominant Western model for thinking about what is part of a human being and for how long has provided an important inroad to seeing that exchanging ubjects interacts with negotiations about what counts as part of the subject. I have added that this attribution of subjecthood is now part of designating the worthy from the unworthy. Establishing who and what constitutes an end, and what comes to be viewed as a plain resource, is of key biopolitical importance. In classifying the sacred from the profane, the worthy from the unworthy, the end from the means, a society makes life and death decisions. Such decisions demand attention. In a so-called bioeconomy, questions about material continuity are no longer ways of determining what will be part of an immortal afterlife: contemporary debates designate the living as persons either receiving or giving ubjects. I therefore depart from Bynum and Rabinow when they refer to material continuity as a set of *enduring ideas*. It not only places far too much emphasis on intellectual history as a driver for social history, it also does little justice to the fundamental differences in the institutions involved in producing medieval scholarly texts and the contemporary medical institutions currently craving for ubjects. Debates about material continuity reflect very different sets of ideas over time and interact with very different institutions and games of power. Also, Bynum's proposition cannot explain why some ubjects become so contested while others rarely spark emotional responses.

Transformations in the moral paradigm of market thinking imply changes in the entitlements of people to their own and others' bodies. It implies a fundamental reconfiguration of who may do what to whom, and as such, it commands careful consideration by policymakers and scholars alike. The first people to acknowledge the complexities of the stakes involved should be the ethnographers moving in the wetlands of real people's real lifeworlds. When Max Weber in his classical essay *Science as a Vocation* invoked a religious metaphor to describe the social scientist's commitment to the world, he made it clear that understanding is different from politically designating what is right and what is wrong.[39] Some scholars seem to like the thrill of the morally abject and they flesh out (excuse the pun) "markets in human body parts" for its grisly emotional effect. In contrast, I believe that if the ethnographers studying ubjects take to heart the Weberian vocation, the descriptions produced will illustrate better the complexity of the issues involved.

Endnotes

1. One obvious exception is of course Waldby and Mitchell's, book *Tissue Economies,* in which they make a brief comment which I will cite in its full length:

   "Tissues that we consider essential to the body's integrity and function—organs, blood, skin, the limbs—are strongly invested with ontological significance, and their loss is a catastrophe for the subject. Tissues that are routinely shed or expelled by the body—hair and nail clippings, nasal secretions, saliva, pus, skin particles, urine, feces, sweat—are either ontologically neutral (hair clippings) or ontologically repugnant (urine, feces, pus), the opposite of self value" (p. 84).

   This treatment has some value to it and resonates also with what I have been arguing about the sense of control. However, it remains too superficial. Nail and hair clippings are known to be central for self-value in some cultures and time periods (think also of Bynum's work on medieval theories of resurrection), just as blood and skin can appear ontologically insignificant in certain situations. As argued in Chap. 3, there is no reason to presume a clear ontological difference between ubjects; they all operate on the border of the self.

2. The debates following the textual turn in ethnography in the 1980s baffled ethnography so that it took decades to reinvigorate anthropology as a theoretical discipline; see Hastrup (1995).

3. For an interesting Nietzsche-inspired analysis of anthropological practice as marked by a self-limiting slave ethics, see Linder (2004). For an attempt of revitalizing the comparative anthropological project in search of more general commonalities, see Bloch (2009).

4. See also Clarke (1995).

5. Landecker (2007).

6. Blumer (1971).

7. In line with the reasoning presented in Chap. 1, only that which is seen as having been part of the body has been included, though many tools, machines, and even clothes could be seen as ubjects ambiguously related to persons. Also, I have not included food on the list, though it can be thought of as ambiguously related to the body of a given person. Consider, for example, how Foucault explored the shift in technologies of self from a focus on food intake as an essential practice to sex in *The History of Sexuality*, volumes two and three (Foucault 1986, 1992).

8. Lakoff (2005:106ff); see also Lane (2007) and Rose (2007).

9. Interesting examples include bone scans (Reventlow et al. 2009), prescription by numbers (Greene 2007), and genetic tests (Taussig et al. 2003).

10. See discussion in Chap. 3 of classical anthropological studies of dangerous bodily products as well as Bynum's work on the concern with nail and hair in medieval theology.

11. Contested notions about ambiguously defined entities have been widely debated in the STS literature on boundary objects (Bowker and Star 1999; Star and Griesemer 1989).

12. http://www.lifegem.com/ (last accessed April 22, 2010) and http://www.koerperwelten.de/ (last accessed April 22, 2010). See also Linke (2005).
13. See, for example, discussion of provocative uses of embryos in art in Morgan (2009).
14. Hecht (2003). Consider also how half a century earlier Jeremy Bentham as one of the most influential utilitarians in the Anglophone world had insisted on being publicly dissected. When Hecht uses the term "belief system," she makes a provocative analogy between science and religion. I do not wish to suggest that secular, scientific beliefs are similar to religion. When in the following I write belief system, I refer to the knowledge-axis of biopower. The rise of a scientific belief system implies a set of practices and ways of approaching what counts as knowledge very different from religious reliance on divine authority. For a comparison of science, magic, and religion as belief systems, see Nader (2006). For alternative studies of how current views of the body influence what can be exchanged, see Sanner (2001) and Schicktanz (2007).
15. In the USA, this latter notion is currently taken to a new thrilling extreme in the cryonics movement described by Romain as consisting of a group that could be seen as contemporary American heirs of the position previously held by the members of SMA: "Cryonics is a particular American social practice, created and taken up by a particular type of American: primarily a small fraction of white male, atheist, Libertarian, middle- and upper-middle-income, computer/engineering 'greeks' who believe passionately in the free market and its ability to support technological progress" (Romain 2010:196). Cryonics supporters fiercely support a personal right to their own body in eternity and freeze it down with the expectation that future technological progress will facilitate continued life.
16. Durkheim (2008:37, original emphasis).
17. See pp 38 and 318.
18. Durkheim (2008:318). For the modern reader, this sounds more like Latour's purification work and the constant overflow of hybrids, and indeed, I happily incorporate Durkheim's point about societal ordering along lines of the sacred and profane into my own framework without feeling obliged toward his delineation of a particular religious realm.
19. Zelizer (1979:44).
20. Seremetakis (1991:47).
21. Edgell et al. (2006) have shown that the least accepted position on faith in the USA is atheism, and they argue that religion serves as a cultural model designating society membership. In American sociology, the Scandinavian countries are occasionally seen as totally irreligious (despite Denmark having a state church), mainly because reference to faith is rarely made in public and not as part of the political life (Zuckerman 2008).
22. As argued in Chap. 2, we should be careful not to think of belief systems as determining the mode of production and exchange. The arrows of causation could in fact be turned around, and rather than claiming that, because they are more religious, Americans can exchange some ubjects more like things, the US preference for dogmatic religion could be ascribed to a need for differentiating the worthy ends from the plain resources exactly because so much is fungible and dependent on merit rather than inherent value. That is, the mode of production and exchange could be seen as promoting a need for religious belief and divine order.
23. Wilkinson (2003); see also Cherry (2005) and Taylor (2005a).
24. Kass (1997).
25. This point draws again on the Foucauldian analysis of the clinic as a mechanism for enrolling the lesser worthy bodies in the care for the more worthy bodies (Foucault 2000; Waldby 2000: Chap. 5). See also Bharadwaj (2008) on the production of "bioavailability."
26. It is often unclear what belongs and does not belong to the medical domain. Sometimes reference to "medicine" can form part of attributing legitimacy to ubject-producing interventions, but it may not always have the desired effect if people do not expect biomedical research to serve their needs. When the Human Genome Diversity Project first began collecting DNA samples, it set out to research ancestry, but when the project was met with opposition, it began highlighting potential medical advantages of knowing human variation. This attempt backfired, however, as activists pointed out that medical research tended to be driven by commercial

incentives and would therefore not be in the interest of the indigenous people from whom samples would be taken (Reardon 2005).

27. Taussig (2009); see also Van der Geest and Finkler (2004) for an introduction to cultural differences in hospital practices.

28. Mulkay (1993).

29. Naomi Pfeffer (forthcoming) has explored in detail the ways in which "waste" has fueled the ubject industry in a century-long pursuit of rationalization through which slaughter houses and morgues have become productive sites of post-vital living. As pointed out also by Michael Thompson in his book *Rubbish Theory*, waste is never a permanent state; it is a state of potentiality (Thompson 1979); see also discussion in Hetherington (2004). Waldby and Mitchell suggest that the category of waste plays a key role in the creation of value by facilitating ownership. This recycling of waste to value is paradigmatic for neo-capitalist ideals of perpetual growth. Waldby and Mitchell make the point through the examples of foreskin from circumcision used for skin grafts and placentas used for cosmetic products; see also Cooper (2007, 2008) and Waldby and Mitchell (2006:115). The category of waste can be trusted no more than the unruly ubject, however, as illustrated also by Pfeffer. That an ubject can be disposed of is not the same as making it reusable without contestation.

30. Bencard (2009), Ehrich et al. (2008), Morgan (2009), Palmer (2009), and Parry (2009). In a more indirect way, Everett (2002, 2007) also describes how her understanding of cells taken from her own son continuously changes making her position herself differently toward their usage and mode of distribution. For a call for using presence and materiality to move beyond textual interpretation, see Gumbrecht (2004).

31. Parry (2008).

32. Green (2008) and Seremetakis (1991).

33. Freud (1978). This and the following quotation are from pp 241 and 230.

34. I would like to thank Omi Tinsley for pressuring me to think more about the role of laughter.

35. Anne Carter has reminded me of Freud's (1999:400) suggestion that those things we cannot bear to confront ourselves with produce a comic effect as a "surplus to be discharged in laugher," that is, a mode of letting a repression go.

36. Bakhtin (1965).

37. Durkheim (2008:412).

38. This interest in ambiguity and also what it produces is also what stimulates the recent volume on *Social Bodies* by Lambert and McDonald (2009).

39. Weber (1947).

References

Bakhtin M (1965) Rabelais and his world. The Massachusetts Institute of Technology Press, Cambridge, MA

Bencard A (2009) Life beyond information: contesting life and body in history and molecular biology. In: Bauer S, Wahlberg A (eds) Contested categories: life sciences in society. Ashgate, Farnham, pp 135–154

Bharadwaj A (2008) Biosociality and biocrossings: encounter with assisted conception and embryonic stem cells in India. In: Gibbon S, Novas C (eds) Biosocialities, genetics and the social sciences: making biologies and identities. Routledge, Oxon, pp 98–116

Bloch M (2009) Truth and sight: generalizing without universalizing. In: Engelke M (ed) The objects of evidence: anthropological approaches to the production of knowledge. Wiley-Blackwell, Oxford, pp 21–30

Blumer H (1971) Social problems as collective behavior. Soc Probl 18(3):298–306

Bowker GC, Star SL (1999) Sorting things out—classification and its consequenses. The MIT Press, Cambridge, MA

Cherry MJ (2005) Kidney for sale by owner: human organs, transplantation, and the market. Georgetown University Press, Washington, DC

Clarke AE (1995) Research materials and reproductive science in the United States, 1910–1940. In: Star SL (ed) Ecologies of knowledge. State University of New York, Albany, pp 183–225

Cooper M (2007) Life, autopoiesis, debt: inventing the bioeconomy. Distinktion 14:25–43

Cooper M (2008) Life as surplus: biotechnology and capitalism in the neoliberal era. University of Washington Press, Seattle

Durkheim E (2008) The elementary forms of the religious life. Dover, Mineola

Edgell P, Gerteis J, Hartmann D (2006) Atheists as "other": moral boundaries and cultural membership in American society. Am Soc Rev 71(April):211–234

Ehrich K, Williams C, Farsides B (2008) The embryo as moral work object: PGD/IVF staff views and experiences. Soc Health Illn 30(5):772–787

Everett M (2002) The social life of genes: privacy, property and the new genetics. Soc Sci Med 56:53–65

Everett M (2007) The "I" in the gene: divided property, fragmented personhood, and the making of a genetic privacy law. Am Ethnol 34(2):375–386

Foucault M (1986) The care of the self. Penguin, London

Foucault M (2000) Klinikkens Fødsel [Birth of the clinic]. Hans Reitzels Forlag, Copenhagen

Freud S (1978) The 'uncanny'. In: Strachey J (ed) The uncanny. Hogarth, London, pp 219–252

Freud S (1999) The Interpretation of dreams. Oxford University Press, New York

Green JW (2008) Beyond the good death: the anthropology of modern dying. University of Pennsylvania Press, Philadelphia

Greene JA (2007) Prescribing by numbers: drugs and the definition of disease. The Johns Hopkins University Press, Baltimore

Gumbrecht HU (2004) Materialities/The nonhermeneutic/Presence: an anecdotal account of epistemological shifts. In: Gumbrecht HU (ed) Production of presence: what meaning cannot convey Standford. Stanford University Press, Palo Alto

Hastrup K (1995) A passage to anthropology: between experience and theory. Routledge, London

Hecht JM (2003) The end of the soul: scientific modernity, atheism, and anthropology in France. Columbia University Press, New York

Hetherington K (2004) Secondhandedness: consumption, disposal, and absent presence. Environ Plann D Soc Spaces 22:157–173

Kass LR (1997) The wisdom of repugnance. New Republic 216(22):17–26

Lakoff A (2005) Pharmaceutical reason: knowledge and value in global psychiatry. Cambridge University Press, Cambridge

Lambert H, McDonald M (2009) Introduction. In: Lambert H, McDonald M (eds) Social bodies. Berghahn books, New York, pp 1–15

Landecker H (2007) Culturing life: how cells became technologies. Harvard University Press, Cambridge

Lane C (2007) Shyness: how normal behavior became a sickness. Yale University Press, New Haven/London

Linder F (2004) Slave ethics and imagining critically applied anthropology in public health research. Med Anthropol 23:329–358

Linke U (2005) Touching the corpse: the unmaking of memory in the body museum. Anthropol Today 21(5):13–19

Morgan LM (2009) Icons of life: a cultural history of human embryos. University of California Press, Berkeley

Mulkay M (1993) Rhetorics of hope and fear in the great embryo debate. Soc Stud Sci 23:721–742

Nader L (2006) The three-cornered constellation—magic, science, and religion revisted. In: Nader L (ed) Naked science—anthropological inquiry into boundaries, power, and knowledge. Routledge, New York, pp 259–275

Palmer C (2009) Human and object, subject and thing: the troublesome nature of human biological material (HBM). In: Bauer S, Wahlberg A (eds) Contested categories, life sciences in society. Routledge, London, pp 15–30

Parry B (2008) Inventing Iris: negotiating the unexpected sptialities of intimacy. Hist Hum Sci 21(4):34–48

Parry B (2009) The afterlife of the slode: exploring emotional attachment to artefactualised bodily traces. Max Planck Institute for the History of Science, Berlin

Pfeffer N (forthcoming) Insider trading. Yale University Press, New Haven

Reardon J (2005) Race to the finish: identity and governance in an age of genomics. Princeton University Press, Princeton

Reventlow SD, Hvas L, Malterud K (2009) Når kropsopfattelsen bliver udformet af medicinsk teknologi—kvinders erfaringer fra knogleskanninger. In: Hvas L, Brodersen J, Hovelius B (eds) Kan sundhedsvæsenet skabe usundhed? Refleksioner fra almen praksis. Månedsskrift for Praktisk Lægegerning, København, pp 201–212

Romain T (2010) Extreme life extension: investing in cryonics for the long, long term. Med Anthropol 29(2):194–215

Rose N (2007) The politics of life itself: biomedicine, power, and subjectivity in the twenty-first century. Princeton University Press, Princeton

Sanner MA (2001) Exchanging spare parts or becoming a new person? People's attitudes toward receiving and donating organs. Soc Sci Med 52:1491–1499

Schicktanz S (2007) Why the way we consider the body matters—reflections on four bioethical perspectives on the human body. Philos Ethics Humanit Med 2:30–41

Seremetakis CN (1991) The last word: women, death, and divination in Inner Mani. University of Chicago Press, Chicago

Star SL, Griesemer JR (1989) Institutional econology, "translations" and boundary objects: amateurs and professionals in Berkeley' museum of vertebrate zoology, 1907–39. Soc Stud Sci 19(3):387–420

Taussig K-S (2009) Ordinary genomes: science, citizenship, and genetic identities. Duke University Press, Durham/London

Taussig K-S, Rapp R, Heath D (2003) Flexible eugenics: technologies of self in the age of genetics. In: Goodman A, Heath D, Lindee S (eds) Genetic nature/culture: anthropology and science beyond the two-culture divide. University of California Press, Berkeley, pp 58–76

Taylor JS (2005) Stakes and kidneys: why markets in human body parts are morally imperative. Ashgate, Hampshire

Thompson M (1979) Rubbish theory: the creation and destruction of value. Oxford University Press, Oxford

Van der Geest S, Finkler K (2004) Hospital ethnography: introduction. Soc Sci Med 59:1995–2001

Waldby C (2000) The visible human project: informatic bodies and posthuman medicine. Routledge, London

Waldby C, Mitchell R (2006) Tissue economies: blood, organs, and cell lines in late capitalism. Duke University Press, London

Weber M (1947) Science as a vocation. In: Gerth HH, Mills CW (eds) From Max Weber: essays in sociology. Oxford University Press, New York, pp 129–156

Wilkinson S (2003) Bodies for sale: ethics and exploitation in the human body trade. Routledge, London

Zelizer VA (1979) Morals and markets: the development of life insurance in the United States. Columbia University Press, New York

Zuckerman P (2008) Society without god: what the least religious nations can tell us about contentment. New York University Press, New York

Conclusion

With this book, I have explored an ongoing reconfiguration of relations between three interrelated domains usually framed as the "body," the "person," and the "market." I made the claim that changes in relations between them imply reconfigurations in each domain, and in particular, I have shown how the intersection between "body" and "market" affects the extent to which an ubject is seen as part of a "person" as well as how such associations affect the entitlements that people can legitimately acquire in an exchange situation. When coined as a "market in human body parts," the intersection between "market" and "body" arouses strong reactions and attracts horror-infused as well as ideology-motivated (almost prophetic) attention. The resulting debate is a melting pot of values related to key capitalist institutions. It remains fundamentally unclear whether such a market exists. As Bronwyn Parry remarks, the phrase is "underwritten by a powerful supposition: that we would know what a market in commodified bodily parts would look like if we saw one."[1] "Markets in human body parts" is an uncanny image, a fusion of what many think should have been kept apart. It is there, and it is not. The uncertainty is a consequence of the fact that markets and bodies are not well-defined entities occasionally mating; they are metaphors used to make sense of complex phenomena. Indeed, there is no such thing as a market in human body parts waiting to be revealed (or realized), just as it makes no sense to suggest that entities originating in bodies are *not* exchanged between commercial actors in pursuit of monetary gain. When market proponents and opponents claim one or the other, they construe images using categories that are ill suited for the phenomena at hand. These images are productive, however, in the sense that they have important performative effects, but they do a poor job of portraying the terms on which ubjects mostly change hands.

People tend to invest very different feelings in ubjects than in those objects for which market theories were primarily developed—and where the theories in many respects do a better job of describing the terms of the exchange (however disliked by many anthropologists, sociologists, and STS scholars). Hecht

K. Hoeyer, *Exchanging Human Bodily Material: Rethinking Bodies and Markets*,
DOI 10.1007/978-94-007-5264-1, © Springer Science+Business Media Dordrecht 2013

suggested that the atheist, materialist, and rationalist members of the Society of Mutual Autopsy attracted the attention they did in nineteenth-century France, not because their views were shared by the general public, but because they represented a "thrilling extreme" laying out positions from which others could navigate in a burgeoning secular society. Similarly, we might see the stark market proponents as well as the most indignant of the market opponents as representing *thrilling extremes*: their depictions are avant-garde and out of sync with the views of the general public as well as the actual modes of ubject exchange. They nevertheless serve important functions as dry spots of clarity in what is really a wetland of hybridity.

Both market proponents and opponents depend on strong claims about the "real" ontological status of ubjects. Market proponents generally suggest seeing ubjects as things belonging to the realm of plain material resources. Market opponents typically suggest seeing ubjects as parts of bodies and therefore part of persons. Both claims are equally untenable. They represent the worldview of the discussants, that is, their categories, stated values, and preferred discourses, but not the lifeworld in which they and we engage the everyday practices of ubject exchange. Despite the many attempts of market proponents to *define* ubjects as objects, every ubject embodies an unruly potential of subjecthood, and ironically, commercial exchange forms do a great job of unleashing it. Conversely, market opponents tend to imagine bodies and ubjects with everlasting qualities. Both positions rely on a vision of a clear boundary between persons and commodities but disagree about where it goes and how it is drawn. Regrettably, or luckily perhaps, no such clear boundaries exist. We are all continuously made and dissolved. Being alive is one long engagement with material flows, and at one point, the ubject's potential for subjecthood is bound to fade away whereby it will fall back "beyond that limit" and form part of the non-human (*non-worthy*) material world. Attempts of capturing the material essence of a person, or to define the *real* status of an ubject, are akin to searching for the departed in the chimney heat.

This book is, as I have already stated, primarily aimed at ethnographers. It takes point of departure in the conviction that to make more adequate descriptions of ubject exchange, we need to provide more attention to ambiguity. By virtue of their methods, ethnographers are confronted with the productivity of the undefined in a very direct sense—unlike scholars working at a safe distance of the people who undertake the actual muddling through that everyday exchange practices always involve. It is tempting to distil clear worldviews using existing categories and perhaps especially to do so in order to arouse a moral effect, but as Nietzsche points out: "no one *lies* as much as the indignant do."[2] If we wish to do justice to moral complexity, we need to handle ambiguity with care.[3] Obviously, there is no standard recipe for how one might go about that.

I have argued here for problem-specific theorizing, and in this case, I have suggested using a slightly strange vocabulary to avoid the pitfalls of claiming something to be clear which is really contested. Where some say markets and others say gift economy, I identify contestations over the nature of the *exchange system*. Where some say property rights and others say *nullius in bonis*, I detect negotiations of *entitlements.* Where some say objects and other say parts of persons' bodies, I spot an *ubject.* All three concepts are construed to focus on relations rather than artifactualized entities. Ubject, the strangest of the terms, is a temporal relation, and I have used it to capture, on the one hand, a potential for subjecthood rarely acknowledged by market proponents and, on the other hand, the fact that this potential is bound to disappear at some point, despite the claims of inherent dignity made by some market opponents. In the everyday practices of ubject exchange, ambiguity delivers a room (as well as an impetus) for action that ethnographers need to appreciate. Ubjects are exchanged, and money is involved, but calling it a market delivers limited insights into actual mechanisms and the values involved in the exchange. What we need to understand better is how the everyday practices of ubject exchange concretely take place, with what implications, and for whom. Good ethnography can deliver important input to current debates by addressing these topics and by reverting from a morally indignant or prophetically propagating image of "markets in human body parts."

What Is at Stake in the Debate About "Markets in Human Body Parts"?

By identifying market thinking as a historically specific discourse and by deconstructing common perceptions of the body, I have arrived at a set of conclusions about what is at stake in the debate usually tagged as "markets in human body parts." These conclusions are, I admit, different from what is typically argued. My point is not that other scholars are wrong or do not know what they really mean but rather that they make their points in discourses that have performative effects not adequately addressed in the dominant framework delivered by the debate. By drawing on framings from market discourses and particular articulations of the body, they shape the social phenomena of ubject exchange in particular ways, and my purpose has been to explore this very process.

First and foremost, I see the debates as arenas for negotiating societal institutions. If we want to understand how societal institutions change, however, we must not limit ourselves to scholarly debates or view negotiations as purely discursive phenomena. Negotiations of institutions take place as ubjects move in hospital basements, between companies, as well as in and out of persons' bodies. Discourses and material flows are coproduced. My central argument throughout has been that by negotiating ubject exchange, people come to designate each other's *relative worthiness,* they settle *entitlements,* and they arrive at particular *understandings of self.* Furthermore, I have argued that these negotiations have implications for the *trust* people invest in, especially, medical institutions, as well as for the way in which exchanges are performed, and perhaps even for the *legitimacy of market thinking itself.* In this section, I will dwell a little more on each of these points before I address the debate more on its own terms with respect to its suggested policy solutions.

Worthiness, Entitlements, and Understandings of Self

My point about *worthiness* drew on an analogy between what Durkheim identified as a religious system and the current demarcation practices discriminating worthy ends from plain resources. References to "markets" have become integrated into a secular belief system surrounding the soulless body (as an heir to a boundary between the sacred and the profane)—probably because "markets" have come to represent total commensurability and a sort of calculative, slightly egocentric set of exchange practices in which it is legitimate *not* to care about others. Not all societies use references to markets to make such distinctions, at least not to the same degree, and indeed, many other ways of distinguishing between worthy ends and plain resources exist and have been in use over time.[4]

In arguing that the current contestation of ubject exchange interacts with understandings of *self,* I build on Bynum's classical work on material continuity. I place it however in a quite different biopolitical framework. If Bynum saw debates about (what I term) ubjects as a recurrent puzzle through which people try to delineate the self, I suggest that this puzzle has very different political implications reflecting the changing institutions involved in different historical periods. I have argued that the reason why references to markets so often spiritualize, subjectify, and/or personify an ubject is because the market institution, partly thanks to the moral paradigm providing market thinking with legitimacy, has come to serve—at least in contemporary secular, capitalist societies—as a marker of relative worthiness. Valuation of worth draws upon phenomenological experiences of love, of respect, even of fear, and such experiences tend to emerge from interaction with willful beings. The moral paradigm of market thinking which separates persons from commodities emerged alongside an emerging humanism reconfiguring worthiness on secular terms. It delivered an institution designating new ways of concentrating power (in the form of capital accumulation) along with new ways of balancing that power (by designating that which could not be bought and sold). The idea that human beings were special and beyond trade can be viewed as an effect of this balancing exercise. In the constant struggle of wills, some are stronger than others, and this special status does not necessarily apply to all human beings, as testified by the surge in trafficking and organ trade. Reflections on material continuity still feed into the semantic attempts of making sense of what is self and not-self, as argued by Bynum, but today, drawing the boundary is also part of designating entitlements to one's own and others' bodies in a quite different sense. It has medical and political implications very different from the theological puzzles of medieval scholars. Ubjects can today be put to other uses than medieval relics, and their exchange takes place on an altogether different scale. Underneath contemporary musings over what an ubject really is now seethes a voracious appetite for bodies. Debates about ubjects therefore have very different implications today than they did in medieval theology.

The motor driving the current harvesting machines is to be found in willful beings wanting more life, better life, longer life. Which ubjects come to be desired (and acquired) depends on an intricate interplay between knowledge, power, and moral legitimacy. What comes to figure as a resource depends on available technologies, forms of access, and normative judgments. There is no such thing as a free-floating *will* wanting a piece of bone, a kidney, or a piece of artery: every will is informed by a medical regime setting this particular ubject as solution for each particular problem. It only makes sense to talk about wanting to "buy an ubject" when societal structures use money as a form of power, when technologies facilitate a transfer, and when longing for boundless life and health is socially legitimate.

Trust

Turning to the implications of current ubject contestations for the trust people invest in medical institutions, I first wish to highlight the importance of the fact that bodies entering medical institutions now potentially fulfill a double role as both resources and as deserving ends. Debates about ubject exchange tend to focus on a separation between givers and receivers, just as I have done above. Nancy Scheper-Hughes, for example, focuses on the unidirectional flow of ubjects from poor to rich, from south to north, from female to male.[5] This type of inequality is important, and it has implications for the trust invested by those most likely to deliver ubjects. However, it should not fully derail attention from the fact that many ubjects are procured from people who themselves undergo procedures to enhance their health. Patients receive a prosthetic device and donate their bone; they undergo tumor surgery and donate research material; they undergo fertility treatment and donate surplus gametes and embryos; they undergo heart bypass and donate stem cells from the surgery leftovers.[6] A third of all human heart valves are donated by people at the receiving end of a full heart transfer.[7] When bodies increasingly occupy such double roles as both means and ends, trust can be challenged in special ways. The biopolitical game of relative worthiness is thereby being played out in the very same body. Patients opt for care from people who simultaneously ask for their help as donors, and they therefore need to balance the risks incurred through saying yes with those potentially incurred by saying no, for example, an impaired relation to the health-care provider. Granted that their bodies occupy this double role as both means and end, how will people know that the doctor does not "take more than supposed to" (borrowing a phrase from a patient in Chap. 4)? Usually, the answer is through strict policies on informed consent. However, in informed consent procedures, patients are invited to share the appetite for ubjects and view their bodies as resources. In a sense, they are made aware of their bodies as a reservoir of desired ubjects and that realization need not have a calming effect.[8]

Studies aligning themselves with an ambition of social critique have a tendency to focus on issues of mistrust rather than reasons for upholding trust. It seems to be a fact, however, that when donors trust the medical institutions taking care of them, ubjects circulate relatively freely and create a minimum of attention. Donating ubjects for medical purposes can be a gracious experience when only donors believe that they will not be exposed to unnecessary risk and that recipients will use the ubjects with care. It is rarely a problem that an ubject is *used*; on the contrary, in a secular belief system, the notion of utility can be helpful in infusing ubjects with meaning facilitating their travels. In organizations where people trust the procurement staff, there is little need for explicit recategorizing the ubjects: the status of the ubjects can remain undefined. As ubjects move into laboratories or freezers, their function and meaning gradually change, but as long as boundary infrastructures separate the respective user groups, nobody is confronted with the gradual changes in valuation.[9]

Price setting, however, involves a very explicit change in valuation and commercial and monetary aspects of ubject exchange therefore elucidate changes in the stakes held by new user groups that could have remained unnoticed within a realm of "public" research. It also indicates a valuation which might overrule respect for the donor's longing for care. Perhaps this is why many of the current ubject exchange systems are characterized by a "public" component blurring the first part of an ubject's journey away from the person in whom it originates. Only at later stages, where the user groups have not met the person in whom the ubject originates, is it given a price. Trust probably partly reflects donors and users not knowing the exact steps in the transition.[10] Even if the critical analyst might be appalled by what people do not know, there is reason to pay serious attention to all the efforts undertaken not to know and to remember how comforting it can be for donors not to be confronted with changes in meaning and valuation of donated ubjects.

Most policies enacted to ensure trust in the medical organizations handling ubjects nevertheless emphasize the need for more information. They produce transparency policies and install informed consent requirements.[11] Trust, however, is not the same as knowing. Breakdowns in communication tend to be investigated as prime reasons for systems failure, but ubject exchange might in fact be an instance where it is important to facilitate cooperation with little or no communication. What we chose not to know is often essential for how organizations work, as argued in Chap. 4.[12] To know exactly how and why an ubject is used is not necessarily comforting knowledge, and perhaps overcommunication is part of creating the very problem later defined as "lack of trust." On the other hand, the strictness many organ procurement organizations exert in terms of hindering contact between donor relatives and recipients ignores that these transplants already involve a lot of drama and awareness of potential new partial connections. Rather than assuming all interest in knowledge about exchange partners to be irrational, facilitating contact in cases where it helps people to make sense of their experiences might prove a better way to ensure trust in the involved medical institutions.[13]

My point is that the route to a higher level of trust is not to be found by focusing only on how much or how little information patients and donors are presented with. It might be found, however, in listening to patients' stated informational needs and, considering the arguments in Chap. 4, in setting up incentive structures that do not invite medical professionals to put patients at risk, infrastructures that do not challenge patients' sense of control, control mechanisms that ensure expected usage of collected ubjects, and research infrastructures designed to address public health needs. I admit that these are all somewhat comprehensive tasks. A place to begin could be to address who can make money out of what. That would include scrutiny of the practices that made it possible, for example, for a publically employed British surgeon to sell bone procured at public hospitals or for American companies to make a profit out of stripping cadavers, as in the Alistair Cooke case. Instead of delivering more information to the donating subject or to the unengaged public (the ever-mentioned "educational campaign"), organizations wanting to ensure trust could commit themselves to close the room for illegal action left open by a willingness to buy without asking questions, and they could begin to develop methods for regulating compensation policies, including the setting of standards for calculation of processing fees and recovery costs.

The Legitimacy of Market Thinking

Finally, if the rise of market thinking is intertwined with the rise of moral ideas about the person (and, per extension, the person's body) as being beyond trade, it is important to consider what the rise of a so-called bioeconomy might imply for ideas about "markets." How will capitalist models change, which new forms of resistance will emerge, and how will such resistance be incorporated into new forms of power? There are high stakes in this battle. Considering how dominated by market thinking contemporary American society happens to be, it is perhaps no coincidence that in particular US scholars feel an urge to pose pro-market arguments at the thrilling extreme of the spectrum. The ways are which the body is constantly set apart from the rest of the material world are of course deeply provocative for adherents to a market ideology. Any hegemonic discourse promotes obsessive work of purification, and for market thinking, I guess it is only natural to enroll the body into the regime supposed to rule the rest of the material world. When economists take market thinking to this extreme and draw supply/demand curves for organs in order to calculate "equilibrium prices,"[14] they might even think that they are stabilizing the current ideology by showing its universality. However, legal endorsement of price-setting mechanisms based on property rights to ubjects does not just extend "market thinking" to a new realm; it involves changing its basic rules of legitimacy. The operative workings of the exchange system are changed when its tenets of legitimacy transmute. If the moral paradigm of market thinking outlined in Chap. 2 reconfigures itself to include the body in the realm of property, new forms of balancing resistance will emerge. I am not sure "stabilization" will be the word best describing the effect of such resistance.

The moral paradigm currently producing special rules for ubject exchange is seen as irrational by some market proponents. It is portrayed as a sort of cultural barrier hindering efficient exchange. It is important however not to view the agency that makes ubjects move in particular ways as somehow "false" or "illusionary." The moral agency is indeed very real, and it has very real effects. It produces policies banning organ and tissue sale, it stimulates innovation in intellectual property rights to compensate companies in alternative ways, and it creates a consistent blindness toward the calculation of processing fees, recovery costs, and compensation schemes. Not only does it produce societies claiming there are no monetary interests where clearly there are, it puts some of the people most dependent on getting hold of money at great risk of losing it. So-called kidney vendors in poor countries operate with no protection and risk losing not only their kidney but also the money they had hoped to make. According to one study, they are on average promised one third more than they are paid, and yet they have nowhere to turn for help when cheated.[15] The protection of persons delivered by the moral paradigm of market thinking in this way favors some persons over others. Basically, market opponents have to face the uneven consequences of their moral judgments. Rather than assuming that answers to the "what-to-do questions" can be found by defining the ontological status of the ubject, I have suggested following the involved actors and exploring their concerns. As argued in Chap. 4, my own hunch is that for most people, what really matters is their phenomenological sense of control and the range of alternative options.

In controversies about ubject exchange, we are negotiating not just ideas about markets but basic ideas about societies and agency. With a successful, though unusual, combination of Bruno Latour and Mary Douglas, STS scholar Maja Horst convincingly suggests that controversies about new medical technologies have become prime scenes for performing what society is and what role human agents play in it.[16] With my invitation to view debates about ubject exchange as sites for enacting change in social institutions, I have followed a similar line of inquiry. Hence, this is what the debate about ubject exchange is also about: what we are as agents and the kind of world in which we live. The first issue includes an additional and quite important sub-question about who is included in the "we."

The Debate on Its Own Terms: Planning the Right Policy?

The debate that I have been analyzing is characterized by a strong urge to prescribe the right thing to do. Especially market proponents talk directly about "prescription for reform"[17] and often aim their books at policymakers. Market opponents are generally less clear in their suggestions and emphasize instead their discontent with existing commercial aspects of ubject exchange. It is now time to return to this debate more in line with its own agenda to briefly discuss some of the most common policy recommendations. The tendency to assume mutually antagonistic positions as either market proponents or opponents illustrates the extent to which market thinking determines what is viewed as available policy options. The first task is therefore for policymakers to free their thinking from this either/or approach.

From Either/Or to Sane Balancing

Market opponents must stop pretending money is not part of the equation when ubjects change hands in complex medical systems. They should engage themselves in *how* it should change hands. I personally think that more emphasis on standards for calculation of compensation and recovery costs would be an excellent place to start. Market proponents, conversely, need to base their arguments on concepts better suited for the analysis of ubject exchange than "market forces" and "invisible hands," as if such fantasies explained which ubjects come to be exchanged and why. The morality underlying market thinking is even more important to confront in order to reach more socially robust policy options. In particular, I think we need to dismiss the common assumption that people are and should be free to use their money exactly as they please. People acquire and accumulate money through state-mediated measures, and the state literally serves as the gun behind the dollar bill, ready to protect "private property." Money is a form of power, and of course, state-sanctioned forms of power need to have state-sanctioned limits.[18]

It can be appealing to think of yourself as self-made and morally entitled to all you can make people do for the money you possess. However, capital accumulation is a social process, and most of the money amassed in those affluent societies currently facing "organ and tissue shortages" is accrued through unequal international exchange. The debate about "markets in human body parts" is a debate about what money, as a particular type of power, can be used for. The task in creating socially robust exchange forms is to balance monetary power in ways that hold wider legitimacy. Once we leave the fortified bastions of the moral landscape, it is easier to appreciate that what the debate tries to settle is how moral rights (associated with monetary wealth) relate to moral obligations (associated with community membership). This is the central policy issue.[19]

Rationality

We often see scholars from the pro-market position describing objections to commercial ubject trade as *irrational* because, as they rightly point out, attempts of setting aside ubjects as forever special and beyond trade are futile. However, even if the specialness of ubjects is not absolute, ubjects do hold a potential making them special, not least because people tend to attach special concerns to them. With ubject exchange, we are dealing with entitlements to, and control of, a bodily space of immense importance to most people (cf. Chap. 3). Objections to ubject trade are not just irrational cultural impediments to sane market forces; they represent quite consistent attempts of retaining control, ensuring trustworthiness, and of angling for care by drawing on available institutions and culturally salient narratives. Even if some market proponents take a thing-like view of ubjects, few of them hold this view consistently. If everything leaving the body (at whatever point it is seen as having left) really were just "things," the natural consequence would be to view as irrational also objections to using cadavers and surgical waste as foodstuffs (just like slaughter animals are sometimes fed the remains of their fellow creatures). Ironically, market proponents rarely take such a thing-like perspective on ubjects. Instead, they propagate the view that donors are entitled to stay related to these "things" and use them to pursue monetary gain. They oppose the abandoned waste rule that used to be at work in hospital settings and wish to reunite donors with ubjects through an ownership relation. They use the potential for a relationship between ubject and person to generate entitlements which in effect set ubjects apart from the rest of the material world. What market proponents suggest is therefore not to treat ubjects as mere things but to use monetary power to ensure desire fulfillment. There are many reasons to be wary about this transfer of mechanisms from manufacture to transplantation medicine. Ubjects are not vacuum cleaners, and their production involves other types of motivation and sacrifice. The moral agency going into attempts of setting ubjects apart from other types of exchange objects informs us about a different set of driving forces than those prescribed by market thinking and constitutes more than irrational opposition. Rational policymaking needs to take well-known social mechanisms (such as emotional affinity for body parts) into consideration when designing new policy solutions.

Implications

The practical effects of providing donors with property rights to their own ubjects constitute a topic which has received a good deal of attention of its own. Legal scholar Carol Rose has pointed out how property rights always seem to engage two different sets of expectations: some assume that they provide the weak with the best means for protecting their interests, others that by making an object into a piece of property, the weak is more likely to lose it.[20]

Similar expectations underlie much of the ubject debate, and again both positions seem to rest on too crude assumptions. The current system, in which donors are not allowed to sell their organs, places in a legal vacuum the weakest and most desperate people who feel forced to sell their kidneys. They have no entitlements and nowhere to turn when cheated.[21] The desperately poor would in that sense therefore stand to gain more leverage by contracting property rights. With their dismissal of property rights, market opponents currently do a better job of protecting middle-class citizens in affluent societies than of protecting the most desperate citizens in poor countries.

Conversely, it must be acknowledged that when market proponents suggest that with property rights the weakest would simply gain an additional source of income potentially helping them out of poverty, it also misrepresents the potential effect of a change of law. If organs and other ubjects were to be defined as property on par with other types of property, they would simultaneously become alienable to the same extent as other material resources. What might begin as monetary debt will potentially reach into people's bodies. In Shakespeare's *Merchant of Venice,* the moneylender Shylock provides a loan bound in a pound of the merchant Antonio's flesh. Should ubjects come to be legally characterized as property on par with other types of property (even while still residing in the body), this will be more than a horrid image from a classic play. It is appealing to present ubject trade merely as an additional source of income—simply adding to available options for the person interested in making the money—but if ubjects can be freely bought and sold, then making them available will come to feature as an obligation when confronted with serious debt. Even with laws stating that moneylenders cannot demand an organ, the social pressure in families and among people indebted to each other will frame ubject sale as an opportunity each individual will feel obliged to consider. The fact that the people already opting for kidney sale tend to do so as a response to debt is very telling in this respect—and so are the consistent reports on moneylenders in poverty-stricken regions who take available kidneys into consideration when providing a loan.[22] Consider also how legitimacy of purchase creates incentives for illegal procurement and supply (as discussed in Chap. 4). Finally, it is worth remembering that when the Office of Technology Assessment under the US Congress in a 1987 report considered providing donors with property rights in response to the growing demand for tissue in the biotech industry, one of the arguments against it was that it would involve liability on behalf of the donor/vendor.[23] With the right to sell follows not only the obligation to sell but even the obligations of a salesman, and in the case of products originating in your body, you will not always (want to) know what you are passing on.

Justice

Besides protection of the weak, the issue of property rights relates to conflicting views about justice. In particular, when discussing research participation, property rights consistently come up as a way of ensuring participants what John Moore is often portrayed as missing: a fair return.[24] Market opponents design various forms of communal benefit-sharing schemes, while market proponents typically opt for individual, financial returns.[25] What makes a fair return for a bodily donation? In Iceland, the parliament opted for a communal deal which ensured the country free access to pharmaceuticals developed by the main pharmaceutical company using the national database. The National Health Services were to benefit financially, not the individual citizen. In my Swedish case, donors opted, not for a personal financial return, but for researchers to feel morally obliged through the gifting of blood: obliged to care for donors and their clinical needs and obliged to direct their research toward medical need rather than financial greed. If the authorities had wanted to honor this second obligation, they would have had to look into the political design of research infrastructures, including research funding regimes and patent systems. They decided to offer an informed consent instead.

Some market proponents find this version of the deal unsatisfactory and suggest paying donors as a way of offering a better return and ensuring donor control. There are many examples of donors who are in horrible situations while the ubjects they donated thrive and do well. The descendants of Henrietta Lacks found it unfair that she contributed cells to develop sophisticated medicine when they could not afford the most basic treatments. Would they as individuals have been able to strike a deal with the laboratory to get access to healthcare, had they or Mrs. Lacks had property rights? I think not. I think the unfairness they face is more deeply entrenched in the premises of the American health-care system than what property rights in one's body would solve. I also realize, of course, that some market proponents do not find lacking access to healthcare unfair at all. But if you think it is unfair that Mrs. Lacks' descendants have no access to healthcare, I do suggest giving up on the idea that property rights in the body would serve as a solution.

Safety

Another topic often discussed in the debate revolves around what creates the safest supply. Ever since Titmuss, safety issues have tended to be linked to incentives for donations, and both sides make equally strong claims about what type of procurement system best ensures safe products and trustworthy donor testimonies—despite obvious failures both in systems using monetary incentives and in the so-called voluntary systems.[26] The time should have come now to look at incentives from a broader social perspective. It need not be a question about whether the donor is offered money or not but what health professionals and potential donors stand to win or lose by following particular screening criteria. In Argentina, a replacement principle in the blood procurement system requests recipients to find donors to replace the blood they use. Ignacio Llovet found that recipients would typically ask close relatives to donate for them. However, when such relatives had exposed themselves to risk, they would tend to refrain from telling the staff in the screening process because they feared the person asking them in the first place would learn about their, for example, homosexual encounters.[27] The disincentive to provide a truthful donor testimony reflected social pressures and control, rather than an interest in monetary gain, and came about almost as a consequence of trying to avoid a "commercial system." I use this example to argue more generally that it is through engagement with the micropractices of screening that safety is to be built into the system, not through adaptation of wholesale for-or-against-market ideologies.

One Body, One Policy?

A final topic for debate is the category of body part as a generic idea and whether or not policies should differ depending on the specific body part and donation situation. Market opponents tend to exclude the body and all of its parts from any kind of trade. Market proponents conversely typically first compartmentalize and build arguments applicable only to quite specific situations and then nevertheless generalize beyond the specific case. Pro-market scholars such as Mark Cherry, James Stacey Taylor, Sally Satel, David Kaserman, and Andy Barnett, for example, base their arguments on kidneys and nevertheless extrapolate their conclusions to "markets in human body parts" in general.

I find both approaches untenable. The way out of the impasse is first and foremost to focus on what seems to be at stake for different actors (in the widest sense of the word), in particular with respect to their sense of control and the range of available alternatives. Here, I am in line with recommendations from the working party headed by Marilyn Strathern under the Nuffield Council, which also suggested a need to accommodate differences depending on the type of ubject and exchange situation.[28]

Problematizations Revisited

The previous section commented on some of the relatively specific recurrent topics in the debate about ubject exchange; I now wish to address the problematizations of subject exchange at a more general level. Problems never come ready-made; they are made to exist in particular ways through labor.[29] A debate names and frames its problems in ways that leave other concerns unaddressed.[30] It dramatizes particular concerns and emphasizes specific solutions until other problems and alternative solutions become unthinkable. It shapes what can be articulated as legitimate concerns in a given social space.

Scarcity

A framing shared by market proponents and opponents is one focusing on *scarcity*. Almost everybody seems to agree that more ubjects are needed; it is just a matter of how to procure them. This focus on scarcity sets the problem in ways that create a platform for discussing alternative procurement solutions and thereby an entry point for market proponents. Scarcity is in fact typically framed by drawing on market metaphors of supply and demand typically used for discussing equilibrium prices. It is remarked that "supplies" are not dwindling, but they are outgrown by a faster growing "demand." Exchange relations have become so much the domain of economic thinking that alternative metaphors are hard to think of, even for bioethics scholars eager to note that ever more ubjects are circulating, though not enough to satisfy the number of people desiring transplants.

Margaret Lock, among others, has challenged this notion of scarcity by pointing to it as a result of expanded perceptions of what is necessary and meaningful treatment. It used to be the case that alcoholism or advanced age, for example, would preclude a liver transplant in the USA, but today, these patients too can be added to the waiting list. The increased number waiting for a kidney partly reflects a rise in renal disease as a consequence of a boom in so-called lifestyle-related diseases that characterize affluent nations as well as the new middle class in developing nations.[31]

As treatment in other areas improves, patients survive until the point when they need a new organ. The demand for bone grows as people of much higher age than previously are offered replacement prosthetic surgery during which bone is used for additive support, and more bone is used as a consequence of new cosmetic procedures, for example lifting cheekbones or widening jaws (there is a certain gender discrepancy in terms of who prefers which). As more and more is deemed treatable, more ubjects are needed. Market opponents have often been integral to this process by (in my view, laudably) emphasizing a right to healthcare. Instead of limiting access to treatment, even market opponents seem to see only one solution: to enhance ubject supplies. However, their main instrument has been "educational campaigns."

From What to Decide to How to Decide

Market proponents use the notion of scarcity to propose a more radical change of the procurement system. On the face of it, they claim to solve the supply problem by enhancing donations through the use of monetary incentives. However, as argued in Chap. 2, they in fact deliver an altogether more far-reaching, and in some ways more discursively robust, solution. They change the very nature of the problem. Political and moral decisions about who deserves treatment, and about who should feel obliged to deliver the needed resources, are transformed through market thinking into a *procedure*—a method for reaching agreement: you get treatment if you can afford it, and you deliver ubjects if you are willing to do it for money. Nobody needs to make a choice on behalf of anybody else; nobody deserves something they cannot get; nobody is made to do what they should not have done. It is well-known in health services research that you cannot remove the waiting list as long as people are entitled to treatment (because more treatment just creates greater demand).[32] Where market opponents need to justify their priorities, the proponents simply need to let "personal desires" be mediated by monetary bargaining power. It is a radical solution to the waiting list problem. In heterogeneous societies, this type of procedural ethics, as we might characterize it, easily stands to win over substantial ethics.[33] Choices come to rest with individuals, not with collectives in need of legitimacy. There are many liberating aspects of this type of reasoning, at least for those who can afford it.[34] However, it is worth noting that it transforms negotiation of moral legitimacy to issues of relative power. It deflates moral agency. If we do want to do good for more than ourselves, we need to counter the supply-demand thinking typical of market thinking in new ways, which will include reconsidering how to lower the amount of people needing organs and other ubjects—for example, by way of structural prevention where determinants of renal disease are dealt with at a societal level.[35]

Valuation

Along with a procedural approach to ethical issues, a particular framing of value is likely to prevail.[36] As discussed in Chap. 3, ubjects hold many types of value, but any given mode of exchange will tend to emphasize some form of valuation over others.[37] In his treatment of gene-patenting history, Richard Gold describes how in US courts, monetary worth has consistently overthrown other types of value assessments. His point is that monetary valuation thereby has framed the types of concerns raised in courts about technology development. With this book, I have provided a long list of examples that complicate any simple claims suggesting that the monetary assessment always and automatically overtakes other valuations. However, Gold's analysis remains relevant by alerting us to the risk of overlooking key concerns, simply because they cannot be articulated in a monetary idiom. There will never be any agreement on what constitutes legitimate concern or who deserves protection. Some would want to include animals into the sphere of compassion and concern, for example, while others see xenotransplants as a "solution" to the ethical problems of procurement from human beings. Basically, however, there is no simple extension of compassion to include everything with a capacity for suffering: moral agency always involves priorities, and someone or something will have to serve as a lesser-worthy means for others to enjoy the privileges of being an end.[38] I think we will continue to need voices from ethics and social science to bring attention to alternative modes of valuation in order to challenge the all too easy conclusions we are likely to reach when our compassion settles on the subjects closest to our hearts.

Needless Death

Market proponents and opponents all seem to agree on a need for treatment. The ultimate problem referred to in all the various problematizations is, however, one for which there is no ultimate solution: death. All the same, the most common framing in support of any particular policy option is to present it as delivering the means for overcoming just that one invincible finality. Remember how, for example, Sally Satel talks about monetary means in donor procurement as the only viable way of avoiding "hundreds of thousands of needless deaths" (see Chap. 1).[39] Certainly, the reference to needless death has great rhetorical strength, and Satel uses this strength quite consciously when she remarks that "the painful reality of needless death translates into all languages."[40] What makes a death "needless" is not all that clear, however, and for a fact, many people around the world die of very plain diseases which are left untreated simply because the affected people cannot afford malaria treatment, penicillin, or AIDS medication. Satel never explicitly argues that the needlessness of a death is to be measured by the monetary bargaining power of the diseased, and I am not even sure that it is what she means. "Needless death" seems more to serve a rhetorical function. After using the expression a couple of times, she emphasizes how unpleasant it can be to feel that you owe people for a kidney that you are not allowed to pay for (she is a recipient herself), and she talks about how unbearable dialysis can be. I do not wish to downplay the suffering that patients on dialysis feel, and I acknowledge also the contours of a "tyranny of gifting," but it is important to understand how "needless death" represents a problematization through which some deaths and some types of life are given more weight than others. And the idea that wealthy people do not need to feel in debt for gifts of health, just because they are wealthy, is sure to meet opposition among the people of fewer means.

The one position that few seem to take in the debate, the almost distasteful position, is to accept finality. With a couple of colleagues, Sharon Kaufman succinctly writes about organ transplants that this "ethical field is characterized by the difficulty, sometimes the perceived impossibility, of saying 'no'—even in late life—to life extending interventions."[41] Why is this so, and what might we learn from confronting ourselves with the unthinkable? It has not always been unthinkable. In *Anna Karenina*, Leo Tolstoy boldly suggested that the main function of a good doctor is to make everybody (including the patient) feel comfortable that they have done what they possibly could without causing too much harm in all the ultimately fruitless attempts of saving the patient. Acceptance of death today represents fatalism or religious dogmatism of a kind which serves almost as an antithesis to everything good that has come about through enlightenment, science, and secular progress. The type of secular research tradition that has provided us with biomedicine (as well as ethics and social science) certainly knows where to place itself in relation to that dichotomy. Still, we might also need to consider new secular answers to basic questions such as: What do you expect to achieve from continued pursuit of treatment?

Bioethicist Daniel Callahan thinks that modern healthcare has to undergo a "culture change" by facing the basic question underlying medical interventions: "How do we decide when enough is enough, and just what might count as 'enough'?"[42] Health-care costs for marginal improvements of health in middle- and high-income groups are accelerating with the tragic consequence that basic health-care delivery is deemed unaffordable for those who could have had significant benefits at little cost. Callahan wants the financial cost to be taken seriously at a societal and not just individual level. In some instances of ubject transfer, money can in fact be saved as people are taken off dialysis thanks to a kidney transplant, but other ubject transfers have little or no measurable health impact. Callahan suggests that by claiming that no priorities should be made when it comes to health, that nothing is too expensive, we are in fact prioritizing just those who are already well-served and sometimes even overtreated. He points to an important problem. The question is when the people being overtreated want to do anything about it.

Instructor Nick Cassavetas' 2009 movie *My Sister's Keeper* might indicate just one instance of demand for a new sort of public narrative focused on when to say no to further treatment. It stages a caring family in which a younger sister was produced as a so-called designer baby to deliver various forms of human biological material for a cancer-stricken older sister. The almost violently loving mother of the two cannot give up the hopes for the ill daughter, and it becomes the younger one's task to end her older sister's misery by calling a court case in defense of her right to her own body. The older sister dies as the younger puts an end to her donations. The movie presents every actor as loving and trying their best; the villain of the story is the inability to accept that life is sometimes unfair, that even young children have to die. The task is to find the right time to give up, without causing too much pain and suffering. The trick of the story, however, is a sentimental illusion providing little guidance outside the world of fiction, namely, that the two sisters (without their mother's knowing) agreed on the right point in time to relinquish further treatment.

To Dream of Remedies Which Are Worse than the Ill

I have discussed how ideas about enlightenment and progress emerged in tandem with a capitalist mode of production. The current manifestations of biomedicine are fusing a utopian desire for boundless health with a capitalist engine claiming to know no limits to perpetual growth. In the technological spaces it produces, there are few institutional mechanisms facilitating a "no" to treatment. As a consequence, we are occasionally confronted with situations bringing to mind Canguilhem's words when he remarked that "to dream of absolute remedies is often to dream of remedies which are worse than the ill" (cf. Chap. 3). Can suffering sometimes be incurred as much through treatment as lack of treatment? After each cured disease follows a new potential cause of death; which one will be worst? What are the secular means of reflection available to the patient searching for the right moment to stop the search for evermore treatment, health, and life?[43] Engaging these issues amounts to a task which easily comes to appear almost ridiculous—simultaneously grave and banal—and frankly, it is today quite a cliché, namely, asking the meaning of life. Paradoxically, that very question is not very meaningful. A longing for life manifests itself in; it cannot be explained or argued. It consists of doing. What can be negotiated, however, is the cost incurred by the people (and animals) supplying the raw materials needed for aggressive treatment regimes. While medical institutions produce statistical numbers on chances of survival, each individual can consider the tragic risk of prolonged suffering. This risk is mostly left behind in the blind angle of the predominant mode of problematizing disease.

All the same, longing for life is fundamentally meaningful, not by way of reasoning, but as a manifestation of will. And what would it all be worth if we could not give into the illusions it produces? The longing for health and survival fosters care when we recognize it in others, and at some point, we all depend on this type of care. When planning social policy, however, it is important to remember that moral agency has a chilling downside. It involves prioritization; it currently fuels an incessant desire for ubjects; it drives a machinery currently destined toward boundless health. Care for some involves sacrifices in others. Therefore, there is no way to be just right with the righteous.

(Ex)changing the Body

With this book, I have entered a wetland and I have come out with no clear prescriptive conclusions. The point was to reach a different position from which to address a morally charged debate, and my hope is more modest than those who want to monopolize the available solutions, namely, that the reader will feel better equipped to decide on his or her own priorities. I have argued that by exchanging ubjects, we are changing bodies as well as social institutions. More specifically, I have sought to provide ethnographers with reflections stemming from my personal engagement with ubjects. I think we can produce more relevant, productive, *and* subversive representations of ubject exchange through case-specific theorizing addressing the particular ambiguities characterizing a given field. The four analytical ambitions which have stimulated this book represent four elements of such case-specific theorizing. I have highlighted the need for paying more attention to moral agency and to avoid conflating moral agency with moral outcome. Secondly, because too much work focuses purely on high-tech medical technologies and makes too broad claims about novelty, I have argued that much is to be gained by relating insights from the research frontier to studies of mundane and well-entrenched ubject-using technologies. Thirdly, I have suggested employing comparative analysis and new partial connections across diverse phenomena to challenge our thinking about what is special in a given setting and in relation to a given ubject. My point has been to avoid ontological assumptions about the specialness of tissue, cells, or bodies. Any ubject can gain importance from ambiguous associations with subjecthood. Fourthly, and most importantly, I hope to have achieved greater appreciation of the productivity of the undefined. A fruitful darkness seethes in the wetlands of ambiguity. Let us enter and see where it takes us.

Endnotes

1. Parry (2008:1134).
2. Nietzsche (2000:229, original emphasis).
3. See also Rapp (1988, 2003).
4. The Durkheimian analogy raises a question about the underlying ontological and political assumptions of my analysis. The social science tradition on which I draw arose in tandem with the described secular-capitalist societal order. The way in which I in Chap. 3 tried to capture a

sense of material agency with the notion of will is not without resemblances to, for example, Henri Bergson's vitalism and other scholars (including Durkheim) who at that time tried to give secular answers to questions about life, longing, and desires. My point is slightly different though. I do not wish to elevate the secular-capitalist order to absolute truth. I accept divinity as a valid explanation for some for the phenomenological experience of worth and willing. Basically, I am not trying to deliver answers to a mystery which continues to create puzzles for people throughout centuries and across social contexts. My interest lies in the political and moral implications of the answers people produce, not in testing their veracity.

5. Scheper-Hughes (2001, 2005). For the gender dimension, see also Biller-Andorno (2002) and Schicktanz et al. (2006). Only in Iran (where donors are offered monetary compensation) do more men than women donate.

6. Pidsley (2010).

7. Jashari et al. (2004).

8. Scheper-Hughes (see above) similarly suggests that people learn to look at each other with a new type of appetite and that this cognitive change precedes exploitive exchanges.

9. Here again I draw on the notion of boundary infrastructure developed in Bowker and Star (1999) and Star and Griesemer (1989). The basic point is that boundary structures serve to separate communities of practice in terms of the meanings they apply to boundary objects moving between them.

10. As argued by Onora O'Neil (2002), it is important not to confuse trust and trustworthiness. The latter can be a good means for achieving the former, but the two need not be related.

11. For some market proponents, information about market terms will settle problems with mistrust; see, for example, Goodwin (2006).

12. Bateson (1972) and Last (1981).

13. Sharp (2006).

14. Kaserman and Barnett (2002).

15. Goyal et al. (2002). See also discussion of donor risks in Lundin (2012).

16. Horst (2003).

17. Kaserman and Barnett (2002).

18. It is common for market proponents to draw an analogy to dangerous professions and suggest a right to expose yourself to danger if you see a point in it, for example, Satel (2008a) and Taylor (2005). However, even here, it is common for states to put limits to the type of danger people can expose themselves to, for example, by introducing work environment rules or demanding drivers to buckle up in the car.

19. See, in particular, Radin's work for a forceful argument along these lines (Radin 1987, 1996; Radin and Sunder 2005) as well as Cahill's analysis of the issues at stake in the stem cell debate (Cahill 1999, 2000). Some market proponents also suggest that better control is needed; cf. the points about standards below (Fabre 2006; Goodwin 2006).

20. Rose (1994).

21. Goyal et al. (2002).

22. Cohen (1999, 2005).

23. US Congress (1987:70).

24. See discussion in de Faria (2009).

25. Cory Hayden (2007) has analyzed the rise of the notion of benefit sharing and its application in research settings and suggests that benefit-sharing programs constitute a particular type of politics, in which benefits as well as risks are privatized to people who in turn must associate in order to negotiate with researchers.

26. See discussion in Berridge (1996), Copeman (2009), and Healy (2006).

27. Personal communication, Copenhagen, June 9, 2010. I would like to thank Ignacio Llovet for allowing me to use his case.

28. Nuffield Council on Bioethics (2011).

29. Foucault (1988).

30. March and Olsen (1976), Shore and Wright (1997), and Spector and Kitsuse (2001).

31. Siminoff and Chillag (1999). For a critical discussion of lifestyle diseases which questions the implicit causality and focus on individual choices which are built into it, see Vallgårda (2011).
32. Kjellberg et al. (2003) and Kjellberg and Søgaard (2004).
33. See also Horst (2008), Kelly (2003), Wolpe and McGee (2003) as examples of analyses emphasizing the power effect of particular framings. From within a philosophical framework, Alistair MacIntyre (1984) discusses the difficulties facing anyone wanting to make ethical claim in normatively heterogeneous societies, and indeed, ethicists such as Norman Daniels (drawing on Rawls) have taken the consequence and focus on procedures for reaching legitimate agreement and not just the substantive content of those agreements (Daniels 1982, 1988).
34. Carol Rose (2005) points out that so-called commoditization often serves politically marginalized populations very well. In the USA, affluent same-sex couples have long enjoyed more extended reproductive opportunities than fellow homosexuals in the otherwise relatively liberal welfare states in Europe. They have been able to buy what Europeans have had to ask for in state-centered systems that have relied on relatively conservative heterosexual norms.
35. Capewell and Graham (2010). Structural prevention aims at changing factors influencing health rather than individual choices. Our current bioethical frameworks tend to focus on individuals, however, and are poorly tuned to societal challenges.
36. I consciously use the subheading valuation to emphasize how value is not inherent to things or the gaze of "valuators," though often portrayed like that in market thinking. As argued by Dewey, valuation is a dialectic process embedded in "action" rather than "being" (Dewey 1947).
37. When ubjects are desired primarily for reasons associated with health, the entitlements of the people donating and using them can be curtailed in as far as they compromise safety. Consider, for example, the conflicts in the UK about families being denied right to collect and use cord blood because they do not live up to national safety standards for stem cell collection (Devine 2010).
38. I here depart with Cary Wolfe (2010), who seems to suggest that an ethics of compassion can address suffering without producing hierarchies of relative worth.
39. Satel (2008b:1).
40. Satel (2008b:2).
41. Kaufman et al. (2009:19).
42. Callahan (2009:3).
43. A lot of ethics scholarship has revolved around this issue, not least in relation to debates about the good death, and yet the answers produced remain detached from the everyday practices conducted in medical institutions in which treatments are pursued (Green 2008).

References

Bateson G (1972) Steps to ecology of mind. University of Chicago Pres, Chicago
Berridge V (1996) AIDS and the gift relationship in the UK. In: Oakley A, Ashton J (eds) The gift relationship: from human blood to social policy. The New Press, New York, pp 15–40
Biller-Andorno N (2002) Gender imbalance in living organ donation. Med Health Care Philos 5:199–204
Bowker GC, Star SL (1999) Sorting things out—classification and its consequenses. The MIT Press, Cambridge, MA
Cahill LS (1999) The new biotech world order. Hastings Center Rep 29(2):45–48
Cahill LS (2000) Social ethics of embryo and stem cell research. Women's Health Issues 10(3):131–135

Callahan D (2009) Taming the beloved beast: how medical technology costs are destroying our health care system. Princeton University Press, Princeton

Capewell S, Graham H (2010) Will cardiovascular disease prevention widen health inequalities? PLoS Med 7(8):1–5

Cohen L (1999) Where it hurts: Indian material for an ethics of organ transplantation. Dædalus 128(4):135–166

Cohen L (2005) Operability, bioavailability, and exception. In: Ong A, Collier S (eds) Global assemblages. Blackwell, Malden, p 79

Copeman J (2009) Veins of devotion: blood donation and religious experience in North India. Rutgers University Press, New Brunswick

Daniels N (1982) Health-care need and distributive justice. In: Cohen M, Nagel T, Scanlon T (eds) Medicine and moral philosophy. Princeton University Press, Princeton, pp 81–114

Daniels N (1988) Justice in health care. Am i my parents' keeper?—An essay on justice between the young and the old. Oxford University Press, Oxford, pp 66–82

de Faria P (2009) Ownership rights in research biobanks: do we need a new kind of 'biological property'? In: Solbakk JH, Holm S, Hofmann B (eds) The ethics og research biobanking. Springer, New York

Devine K (2010) Cord blood in the car park: cord blood stem cells, contamination and DIY collection exposed. Bionews, p 550

Dewey J (1947) Theory of valuation. The University of Chicago Press, Chicago

Fabre C (2006) Whose body is it anyway? Clarendon, Oxford

Foucault M (1988) On problematization. Hist Present 4(4):16–17

Goodwin M (2006) Black markets: the supply and demand of body parts. Cambridge University Press, New York

Goyal M, Mehta RL, Schniederman LJ, Sehgal AR (2002) Economic and health consequences of selling a kidney in India. J Am Med Assoc 288:1589–1593

Green JW (2008) Beyond the good death: the anthropology of modern dying. University of Pennsylvania Press, Philadelphia

Hayden C (2007) Taking as giving: bioscience, exchange, and the politics of benefit-sharing. Soc Stud Sci 37(5):729–758

Healy K (2006) Last best gift: altruism and the market for human blood and organs. The University of Chicago Press, Chicago

Horst M (2003) Controversy and collectivity—articulations of social and natural order in mass mediated representations of biotechnology. Copenhagen Business School, Copenhagen

Horst M (2008) The laboratory of public debate: understanding the acceptability of stem cell research. Sci Public Policy 35(3):197–205

Jashari R, Van Hoeck B, Tabaku M, Vanderkelen A (2004) Banking of the human heart valves and the arteries at the European homograft bank (EHB)—overview of a 14-year activity in this international association in Brussels. Cell Tissue Bank 5:239–251

Kaserman DL, Barnett AH (2002) The U.S. organ procurement system: a prescription for reform. The AEI Press, Washington, DC

Kaufman SR, Russ AJ, Shim JK (2009) Aged bodies and kinship matters: the ethical field of kidney transplant. In: Lambert H, McDonald M (eds) Social bodies. Berghahn Books, New York, pp 17–46

Kelly S (2003) Public bioethics and publics: consensus, boundaries, and participation in biomedical science policy. Sci Technol Hum Values 28(3):339–364

Kjellberg J, Søgaard J (2004) Ventelister—hvorfor opstår de, og hvad kan der gøres ved dem? Ugeskrift for Læger 166(47):4237–4240

Kjellberg J, Herbild L, Svenning AR (2003) Patientsammensætning, aktivitetsændringer og ventelister: rapportering af første modul af et sundhedsøkonomisk forskningsprojektom 1½ mia. puljens effekter på ventelister og ventetider. DSI Institut for Sundhedsvæsen, København

Last M (1981) The importance of knowing about not knowing. Soc Sci Med 15B:387–392

Lundin S (2012) Organ economy: organ trafficking in Moldova and Israel. Public Underst Sci 21(2): 226–291

MacIntyre A (1984) After virtue. University of Notre Dame Press, Notre Dame

March JG, Olsen JP (1976) Organizational choice under ambiguity. In: March JG, Olsen JP (eds) Ambiguity and choice in organizations. Universitetsforlaget, Oslo, pp 10–23

Nietzsche F (2000) Beyond good and evil: prelude to a philosophy of the future. In: Kaufmann W (ed) Basic writings of Nietzsche. Random House, New York, pp 179–435

Nuffield Council on Bioethics (2011) Human bodies: donation for medicine and research. Nuffield Council on Bioethics, London, pp 1–254

O'Neill O (2002) Autonomy and trust in bioethics. Cambridge University Press, Cambridge

Parry B (2008) Entangled exchange: reconceptualising the characterisation and practice of bodily commodification. Geoforum 39:1133–1144

Pidsley R (2010) Heart bypass "leftovers" yield stem cells. Bionews, p 556

Radin MJ (1987) Market-inalienability. Harward Law Rev 100(8):1849–1937

Radin MJ (1996) Contested commodities. Harvard University Press, Cambridge

Radin MJ, Sunder M (2005) The subject and object of commodification. In: Ertman MM, Williams JC (eds) Rethinking commodification: cases and readings in law and culture. New York University Press, New York, pp 8–29

Rapp R (1988) Moral pioneers: women, men and fetuses on a frontier of reproductive technology. In: Baruch E, D'Adamo A, Seager J (eds) Embryos, ethics, and women's rights: exploring the new reproductive technologies. Haworth Press, New York City

Rapp R (2003) Cell life and death, child life and death: genomic horizons, genetic diseases, family stories. In: Franklin S, Lock M (eds) Remaking life and death: toward an anthropology of the biosciences. School of American Research Press & James Currey, Santa Fe, pp 129–164

Rose CM (1994) Property and persuasion: essays on the history, theory and rhetoric of ownership. Westview Press, Boulder/Oxford

Rose CM (2005) Afterword: whither commodification. In: Ertman M, Williams J (eds) Rethinking commodification: cases and readings in law and culture. New York University Press, New York

Satel S (2008a) Concerns about human dignity and commodification. In: Satel S (ed) When altruism isn't enough: the case for compensating kidney donors. The AEI Press, Washington, DC, pp 63–78

Satel S (2008b) Introduction. In: Satel S (ed) When altruism isn't enough: the case for compensating kidney donors. The AEI Press, Washington, DC, pp 1–10

Scheper-Hughes N (2001) Commodity fetishism in organ trafficking. Body Soc 7(2–3):31–62

Scheper-Hughes N (2005) The last commodity: post-human ethics and the global traffic in "fresh" organs. In: Ong A, Collier SJ (eds) Global assemblages: technology, politics, and ethics as anthropological problems. Blackwell Publishing, Oxford, pp 145–168

Schicktanz S, Rieger JW, Lüttenberg B (2006) Geschlechterunterschiede bei der Lebendnierentransplantation: Ein Vergleich bei globalen, mitteleuropäischen und deutschen Daten und deren ethische Relevanz. Transplantationsmedizin 18:83–90

Sharp LA (2006) Strange harvest: organ transplants, denatured bodies, and the transformed self. University of California Press, Los Angeles

Shore C, Wright S (1997) Policy: a new field of anthropology. In: Shore C, Wright S (eds) Anthropology of policy: critical perspectives on governance and power. Routledge, London, pp 3–42

Siminoff LA, Chillag K (1999) The fallacy of the "gift of life". Hastings Cent Rep 29(6):34–41

Spector M, Kitsuse J (2001) Constructing social problems. Transaction Publishers, New Brunswick/London

Star SL, Griesemer JR (1989) Institutional econology, "translations" and boundary objects: amateurs and professionals in Berkeley' museum of vertebrate zoology, 1907–39. Soc Stud Sci 19(3):387–420

Taylor JS (2005) Stakes and kidneys: why markets in human body parts are morally imperative. Ashgate, Hampshire

U.S. Congress, Office of Technology Assessment (1987) New development in biotechnology: ownership og human tissue and cells. Congress of the United States, Office of Technology Assessment, Washington, DC

Vallgårda S (2011) Why the concept 'lifestyle diseases' should be avoided. Scand J Public Health 39:773–775

Wolfe C (2010) What is posthumanism? University of Minnesota Press, Minneapolis

Wolpe PR, McGee G (2003) "Expert bioethics" as professional discourse: the case of stem cells. In: McGee G (ed) Pragmatic bioethics. The MIT Press, Cambridge, MA, pp 181–191

Index

K. Hoeyer, *Exchanging Human Bodily Material: Rethinking Bodies and Markets*, DOI 10.1007/978-94-007-5264-1, © Springer Science+Business Media Dordrecht 2013

Printed by Printforce, the Netherlands